Microscopic Examination of the Ore Minerals

W Myron Davy, C Mason Farnham

Expand your knowledge.
Increase your power.

BCR's Shelf2Life program grew from a strong desire to help libraries share their vital collections with new audiences. BCR is the nation's oldest and most established library network, and their mission is to bring libraries together for greater success by expanding their knowledge, reach and power. By helping libraries digitize and widen access to their collections, BCR's Shelf2Life program helps libraries increase the visibility, use and recognition of their important collections. In addition, the program is designed to help researchers, collectors and all curious readers by putting these editions within easy reach: in print-on-demand or electronic format.

UniversityLibraries
University of Colorado at Boulder

The University of Colorado at Boulder's University Libraries has a threefold mission. First, to be central to the University community's discovery, communication and use of knowledge by providing materials, information and services that support the University's mission. Second, to serve as a research resource for Colorado residents through support for individual, business, cultural, educational, governmental and other information needs. Third, to share resources with the national and international higher education community. In order to continue to meet this mission and their high level of service to the University, research, higher education, local, national and international communities, the CU Boulder University Libraries continually strive to innovate and adapt their collections and delivery options. By participating in the Shelf2Life Program, the University Libraries provide their faculty, staff, students and researchers access to this important content in ways they have never been able to in the past.

MICROSCOPIC EXAMINATION
OF
THE ORE MINERALS

BY
W. MYRON DAVY
AND
C. MASON FARNHAM

FIRST EDITION

McGRAW-HILL BOOK COMPANY, INC.
NEW YORK: 239 WEST 39TH STREET
LONDON: 6 & 8 BOUVERIE ST., E. C. 4

THE MAPLE PRESS YORK PA

INTRODUCTION

The use of the reflecting microscope as a means of determining the identity, relationship, and significance of opaque minerals in ore deposits has been steadily increasing since William Campbell,[1] in 1906, first applied the methods then used by metallographers in the study of metals, to the examination of opaque minerals. Nearly a century ago Berzelius published the results of polishing a specimen of pyrrhotite and suggested the possibilities of examining opaque minerals in this way, but no practical methods resulted from his observations at that time. Even after Campbell's paper, mining and geological literature did not reflect any great interest in the subject until six or seven years had passed, when some admirable papers described the studies on particular types of ores. About this time the laboratories of mining geology at the Massachusetts Institute of Technology, Harvard University, and Leland Stanford University, as well as a few other investigators, were carrying on extensive researches in this field of study. In 1916, Dr. Joseph Murdoch[2] published the results of numerous microchemical tests, arranging them in a determinative table which was by far the most complete work on identification of minerals under the metallographical microscope, and his book deserves the credit for much valuable pioneer work. Several years have now passed and new ideas have originated as the result of further study along this line.

Work carried on steadily in the laboratories at the Massachusetts Institute of Technology on a wide range of ores, and at Harvard University on the original collection of minerals that formed the basis of Dr. Murdoch's book, justifies the following conclusions:

1. Fine distinctions in color value between the many so-called white minerals cannot be depended upon as a safe property on which to base the major classifications in making identity determinations. It has been found that hardly any two persons can

[1] CAMPBELL, W. "The Microscopic Examination of Opaque Minerals." *Economic Geology,* Vol. 1, 1906, p. 751.

[2] MURDOCH, J. "Microscopical Determination of the Opaque Minerals." Preface by L. C. GRATON, John Wiley & Sons. 1916.

agree in the application of such terms as bluish white, pinkish white, creamy white, purplish white, etc. However, if two microscopes and a comparison eyepiece are used, or if a standard mineral can be mounted adjacent to the unknown, most observers will agree in pronouncing the unknown to be lighter, darker, or the same as the standard; but at best this method is awkward, especially when the unknown mineral does not occur on the edge of the section. Some minerals change color after polishing, the kind of light used effects the colors seen, and a few cases have been observed where two freshly polished specimens of the same mineral differed so greatly in color as to fall into widely separated groups. The observed color of a mineral may be greatly influenced by the colors of other minerals in the field, for example, chalcopyrite is yellow when seen alone, but if seen adjacent to native copper it will appear to be a decided olive green. For these reasons the scheme of identification used in this book is independent of color distinctions which are only used in conjunction with many other properties as an ultimate means of separating the different minerals that fall into a final group.

2. Detailed descriptions of the way in which microchemical reactions proceed are in many cases valueless because of the fact that two specimens of the same mineral seldom yield exactly identical results. In nearly every case, however, they will check as far as reacting positively or negatively with a reagent is concerned, which is all that is needed for identification. Moreover, it has been found that slight differences in the character of the mineral surface, slight differences in the concentrations of the reagents, and sometimes even the crystallographic orientation of the polished section have an important influence upon the reactions.

3. The number of reagents which can be used advantageously throughout a determinative scheme is limited to four or five. Many others have been tried, but their application is very limited and as time goes on the tendency is more and more toward simplification. In this work no attempt is made to list all tests known to date, but only those supplementary ones are included which may be valuable in differentiating between two or more minerals falling in one final group. As emphasized in the following paragraph, in most cases an easily applied blowpipe test is convenient and adds far more confidence to the determination.

4. As in the early application of new developments in all lines, mineragraphy has been pushed beyond its natural limitations in displacing other determinative methods and now a slight reaction is setting in. This art gives us our most useful means for examining into the genesis of ore deposits, and is in itself a very valuable method of mineral determination; in certain cases, the best known method. But in the determination of the minerals in an average polished section of ore it is a handicap to restrict the examination to mineragraphic tests alone. One of a group of possible minerals can usually be run down quickly and confidently by gouging out from the edge of the section a small piece seen to contain the unknown, and treating it on charcoal before the blowpipe, or heating it in the open or closed tube, etc. The objection has been raised that the unknown cannot be cleanly separated, but very often all the other minerals are known and the reactions yielded by them can usually be discounted. There is never any need to run through a blowpipe analysis as the microchemical reactions narrow the possibilities down to a few minerals, and one or two tests suffice. Consequently it has been found exceedingly helpful to include the characteristic blowpipe reactions for each mineral along with its other properties in the tables.

The foregoing considerations have largely influenced the arrangement of the tables and the choice of the material to be included in them.

The microchemical reactions listed are from various sources. The authors have independently studied one hundred and forty-three mineral species, thus covering all except a few of the rarest varieties; and as most of these species are also listed in Dr. Murdoch's book, it is believed that this triple check results in the accurate determination of the reactions.

Fourteen minerals not previously described in mineragraphy have been tested and included in this work. A few very rare and doubtful species and some mixtures described by Murdoch have been omitted as it is felt that their presence in the tables complicates and adds little of practical value for the average user. The impression must not be gathered, however, that the so-called rare minerals are unimportant for, although only the more common minerals are encountered in nine out of ten ores examined, it is of great importance to recognize the rarer ones when they do occur. Furthermore, these uncommon minerals are sometimes found in small quantities or as fine intergrowths and their presence is never suspected until examined microscopically with vertical illumination; consequently they become less rare and take on a greater importance in mineragraphic study than has been the case with former mineralogical methods.

Mineragraphy is used throughout the book to designate the art of polishing, identifying, and examining the ore minerals under the metallographic microscope. *Minera* is late Latin meaning *ore* and *graphy* has been borrowed from the well established *metallography*. Whitehead[1] first used this name in print, and although mineragraphy as a word is not altogether pleasing, a name is much needed and it is here used with the hope that something better will be suggested.

It is believed that an infallible determinative table has never been devised on any subject, and the task is even more difficult in this case where the available material is at best new and little-tried. The beginner, of course, must acquire familiarity with both the tables and the general appearance under the microscope of the more common minerals, and even the experienced mineralogist must serve a short apprenticeship in order to obtain good results. In preparing this book it was felt that a text was needed which could be used by both the profession and the student of mining and geology. Communications addressed to the laboratories of economic geology of the Massachusetts Institute of Technology indicate that the study of ore minerals in reflected light is to be introduced into the curriculum of several more of the country's larger institutions, and that a statement of the latest practice was desired. It is hoped that this rather brief review of the subject will serve such a practical purpose until such time as the combined experience and discoveries of many workers greatly increase our present knowledge along this line.

The authors wish to express their appreciation to Professor Waldemar Lindgren whose able and patient guidance has made the book possible, and to Professor L. C. Graton whose keen interest in this work has ever been a helpful inspiration.

[1] WHITEHEAD, W. L. "Notes on the Technique of Mineragraphy." *Economic Geology*, Vol. xii, No. 8, 1917, p. 697.

TABLE OF CONTENTS

MICROSCOPIC EXAMINATION

OF

THE ORE MINERALS

CHAPTER I

TECHNIQUE OF POLISHING AND EXAMINING THE SPECIMEN

Preparation of the Specimen

The successful determination of the identity of minerals with the reflecting microscope depends primarily upon having all minerals in the section properly polished with little relief, free from scratches, pits, and vugs; and without having changed the chemical character of delicate sulphides during the grinding process. It has been observed, for example, that chalcopyrite if ground under pressure on a rapidly revolving dry lap will develop bornite and chalcocite; and that pyrite when treated likewise will develop limonite.

In 1917 W. L. Whitehead[1] pointed out the fact that mineragraphic sections require grinding and polishing treatment fundamentally different from that of metals for metallographic study. A metal specimen is usually one of uniform hardness and possesses far more ductility and toughness than an ore section which often consists of several minerals of widely differing hardness and of more or less brittleness. Pyrite, a mineral harder than almost any metal, is so universally present in ores that it often must be polished in the same section with a mineral so soft as to be scratched by the finger nail. When purely metallographic methods are used in polishing ores this difference in hardness presents formidable difficulties to good work. The soft broadcloth surface of the wheel swells outward and rapidly wears down the soft minerals, developing a relief so pronounced that it is

[1] WHITEHEAD, W. L. "Notes on the Technique of Mineragraphy." *Economic Geology*, Vol. xii, No. 8, 1917, p. 697.

impossible to focus upon two adjacent soft and hard minerals, especially at high magnifications.

In the laboratory, specimens are usually desired for two distinct purposes; first, for visual examination and second, for photomicrographs. Comparatively little effort is required in polishing sections for the former purpose, and while a badly scratched and pitted surface should never be tolerated, very fine and shallow scratches do not interfere with visual study at any magnification. However, these comparatively minute scratches and pits make photography difficult, and they require exhaustive and careful polishing for their elimination.

The following procedure is in use at the Massachusetts Institute of Technology and with minor variations can easily be adapted to any grinding and polishing equipment. First, a portion of the specimen is chosen which appears to represent all the important features present, an area of a square inch or less usually being sufficient. This may be carefully chipped or cut off with a diamond saw. The latter method is by far the best when such equipment is available, as it rapidly yields an approximately plane surface at any desired orientation and the opposite face of the cut can be made into a thin section if required. The saw used is 10 inches in diameter and is driven at 1600 R.P.M.

Any large inequalities in the surface chosen should be removed by grinding wet upon a steel lap wheel with 150 emery or alundum. A 10-inch horizontal wheel revolving at about 200 or 225 R.P.M. is used. After thorough washing of the hands and specimen the grinding is continued on an 8-inch horizontal glass wheel driven at 1800 R.P.M. and using the finest optical alundum (manufactured by Norton & Co., Worcester, Mass.) mixed to a very thin slime with water. This is perhaps the most important stage in the preparation. Considerable pressure may be exerted at first to quickly remove the scratches caused by the coarse abrasive; then grinding should be continued with light pressure until the surface is free of all except very fine scratches and pits. More than a minute or two is seldom required at this stage, but when needed more time should be taken, as a minute spent here saves ten minutes later.

The true grinding part of the operation is completed and from now on the purpose of the work is to obtain a final polish. The same preparation of finest optical alundum is next used on a high speed 8-inch wheel covered with tightly stretched coarse linen (about 15 threads per cm.). This wheel is horizontal and is driven at 1800 R.P.M. The specimen should be lightly held and constantly turned for 30 to 60 seconds, finishing up with about 30 seconds polishing without applying fresh alundum. The abrasive breaks down and developes a fine polish as seen by the naked eye.

This polish is improved by holding lightly for 30 or 60 seconds on a similar wheel covered with fine linen (30 threads to the cm.) using a slime of rouge and water. This wheel is an 8-inch horizontal lap wheel driven at 1000 R.P.M. After thorough washing and drying the final surface is obtained on a wheel similar to the preceding, but covered with tightly stretched calf skin. A split calf skin with the grease removed and finished with a very short nap on the flesh side is used. Care should be taken here to prevent overheating with consequent altering of the sulphides and burning of the leather. With a little practice 10 minutes will suffice for the entire process of cutting and polishing a surface comparatively free from scratches and troublesome relief, which can be satisfactorily studied even under the highest magnifications.

Out of a number of specimens studied from one deposit, two or three will usually be chosen for photographing, and these should receive additional preparation to eliminate the very fine scratches which cannot be avoided with the foregoing procedure alone. Chromic oxide has been found to yield the best results of all materials tried to date. Its use is introduced between the second optical alundum and the rouge wheels. The finest commercial chrome oxide is levigated to remove any coarse material and is used on a coarse linen wheel revolving at 1800 R.P.M. or faster. Careful treatment here eliminates the fine scratches produced by the optical alundum. The specimen, as before, is finished on the rouge wheel.

In discussing a method of polishing ore specimens no fixed rule can be formulated and the more work that is done the more evident it becomes that each type of ore requires specialized treatment to obtain the best surface for photography. On the other hand, as previously stated, only a small portion of the specimens studied need be so prepared and the standardized and rapid method first described produces excellent results when only visual studies are to be made.

The quality of the illustrations in many of the recent papers on mineragraphic study proves that some of the laboratories have highly developed polishing methods and criticisms and suggestions from them should advance our knowledge of the subject.

FIELD PREPARATION OF A SPECIMEN BY HAND

The engineer studying or developing a prospect, or the geologist in the field usually lacks the necessary equipment as above described, yet it is often important and helpful to clear up certain points without waiting until the apparatus is available. In this case fairly satisfactory results can be obtained by hand with a supply of coarse and fine emery or alundum, rouge and leather. They should all be used on surfaces similar to those described under mechanical polishing methods. So valuable does this line of investigation promise to become for the examining engineer and geologist and so inexpensive is the microscopic equipment necessary that a supply of the requisite materials should be a part of every field outfit.

EXAMINATION OF THE SPECIMEN

A good microscope in common use is the Sauveur and Boyleston metallographic microscope manufactured by the Bausch and Lomb Optical Company, and illustrated in Fig. 1. It is very convenient for examination in daylight and with moderate magnifications. The objectives usually used are of the short mounted type and are of 16 millimeters and 4 millimeters focal length. Many types of microscopes, however, can be fitted with a vertical illuminating prism and be used for ordinary mineragraphic purposes. The simple optical principle involved is illustrated in Fig. 1. Light enters at and is reflected vertically downward by the prism, strikes the polished surface, and is reflected vertically upward. A thin plane glass illuminator is used in place of the prism at higher magnifications and with artificial illumination, as it gives a sharper image. The plane glass reflector, however, is unsuited for use with daylight as it does not give sufficient intensity. Recently a silvered reflector to replace the prism has been placed on the market and has proved satisfactory with all kinds of light.

A devise that gives the location of any particular feature in a polished section, and enables this feature to be brought again immediately into the field of the microscope, may be easily constructed in a few minutes. (See Fig. 2.) A piece of metric cross-section paper ruled to millimeters is sealed between

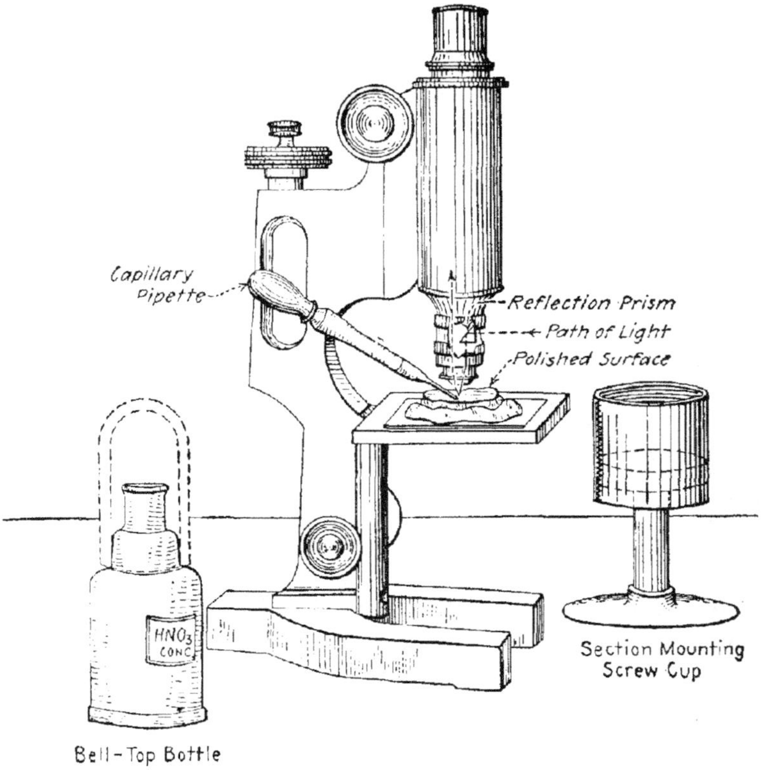

FIG. 1.

two thin pieces of glass which are about four and one-half centimeters long and two and one-half centimeters wide. (Glass slides used in mounting thin sections may be used.) To the under

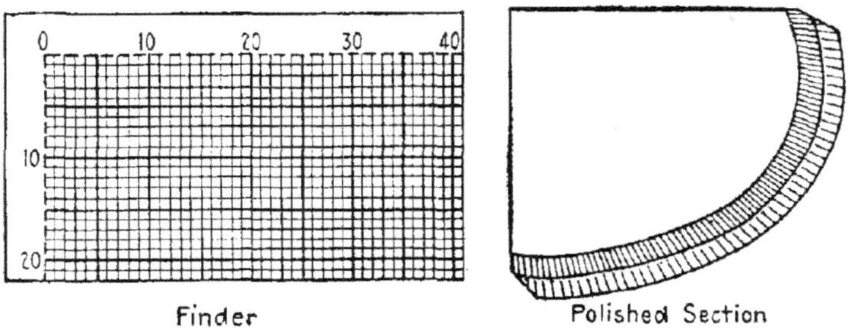

FIG. 2.

side, two narrow strips of wood are glued to two adjacent edges so that their intersection will form a perfect right-angled corner. The cross-section lines should be considered coördinate lines and the point directly above the intersection of the wooden strips be considered the 0-0 coördinate. It is necessary that two sides of the polished section be ground so as to form a right-angled corner. This rough grinding requires only a few minutes additional work on the section, and the edges thus produced allow all minerals at the edges to be compared directly with standard minerals. In subsequent polishing, these sharp edges will not injure the polishing cloth, provided the corner is pointed in the direction of revolution of the lap. When any feature in a polished section is to be located, the procedure is as follows: (1) raise the objective about a centimeter and gently place the finder upon the polished section, carefully fitting the corner of the section into the corner formed by the wooden strips on the finder; (2) focus upon the glass surface of the finder, using the 16-mm. objective, and place a minute drop of ink with a fountain pen in the center of the field; and (3) remove the finder and read directly the location as x millimeters to the right of, and y millimeters down from the coördinate 0-0. Relocation is made in reverse order. This method has been used on a large number of sections, and though rather crude, has been found to be simple and effective, and almost indispensable with the highest magnifications.

The specimen is mounted by being pressed into a lump of modelling clay on a glass slide or similar surface. The polished surface must be perpendicular to the axis of the microscope and probably the most convenient method of leveling it accurately is by means of a screw bottom cup also illustrated in Fig. 1. A piece of 2-inch pipe about 2 inches long is threaded on the inner side and fitted with a threaded plug bottom. The slide with the specimen pushed approximately level into the clay is placed inverted across the top of the cup. By adjusting the bottom the slide is made to just clear the rim and it is then pushed firmly down. The surface of the specimen is now parallel to the slide which will rest upon the stage and consequently will be perpendicular to the axis of the microscope.

For preliminary study and when applying reagents the 16 millimeter (low power) objective is always used. Focus upon the surface to be examined and adjust the vertical illuminator until the surface appears brightest. The opaque minerals will appear white or colored, while the transparent gangue minerals will be black or very dark gray. Tests for hardness, sectility, color of powder, chemical behavior, etc., may now be applied.

A fine sharp needle mounted in a small handle five or six inches long will serve to test for hardness or similar properties. The user should accustom himself to always holding it in about the same manner when testing for hardness in order to correctly compare results. Those minerals that scratch easily without pressure or with very light pressure are classified as of low hardness; those which scratch very slightly with light pressure or easily with heavy pressure are classified as of medium hardness; and those minerals which scratch very slightly or not at all with heavy pressure are classified as of high hardness. The opaque minerals range through a uniform gradation from the softest to the hardest, but after a short time it is not difficult to classify most of them readily. When a mineral is on the border line between two degrees of hardness it is usually found in both positions in the tables.

Sectility or brittleness determination is another convenient test which is sometimes of value. When testing for this property the needle is pushed across the mineral surface with a rotary motion; when the mineral is brittle the powder formed, if any, usually flies away from the edge of the scratch; if the mineral is sectile, little or no powder is formed and the needle penetrates more or less as in cheese.

When a heavy scratch is made by pushing the needle point across the mineral surface the color of the raised edge of the furrow or of the powder formed is very characteristic in the case of certain minerals and this property is an important aid in identification in these cases. See Table 9, page 122.

The microchemical tests are by far the most important in identifying the minerals and should consequently be applied with care and uniformity. The reagents are conveniently applied to the polished surface by the use of a pipette with a fine capillary opening fitted with a small rubber bulb. With such a fine pipette a very small drop of the reagent can be applied exactly where desired and great nicety of manipulation is easily possible. It has been found desirable to use a bell top bottle in which the pipette fits as a ground glass stopper, as the microscope is thus guarded from the acid fumes. A method of applying the reagents described by E. S. Bastin[1] has been found convenient in some cases, especially with the weaker reagents. In this method a small strip of blotting paper tapered at one end is used

[1] *Economic Geology,* Vol. xi, No. 7, 1916, p. 691.

for touching the surface with the reagent and another clean strip is used for removing it to stop the reaction when desired.

The tests should be watched under the microscope at all stages and the following points observed: (1) Is any effervescence produced? (2) Does the surface change color? (3) Is there any structure developed (cracks or cleavage, etc., made more prominent)? (4) Do the fumes of the reagent affect the mineral surface beyond the edges of the liquid? After the reaction has proceeded for ten seconds or up to a minute, depending upon rate of attack, the specimen may be removed from the stage and washed by directing a fine stream of water on the surface from a small wash bottle or dropping tube. The section is again examined under the microscope without rubbing and the following points noted: (1) Is the area covered by the liquid tarnished? (2) Is there any structure or etching developed in this area? (3) Is the area subjected to the fumes tarnished? The specimen is again removed from the stage, carefully dried, and rubbed lightly on a block covered with chamois skin, and again examined to determine whether or not the effects are persistent. Before rubbing, the surface is sometimes slightly dirty due to the evaporation of some of the liquid even though the reaction is entirely negative. This should not be mistaken for a slight positive reaction and may usually be avoided by carefully blotting the washed surface with a soft handkerchief or cloth.

Electrical conductivity, in many instances, may serve to distinguish minerals of similar physical and microchemical properties. The test for electrical conductivity is made under the microscope, using a 16-mm. objective, and the equipment must be adapted to mineral surfaces as small as two millimeters in diameter. Therefore steel needles had to be used for contacts. At first it was thought that these contacts would have to be placed a fixed distance apart in order to give constant readings, but experiment has shown that the difference between having the terminals a centimeter apart and very close together, is too small to be recorded. The apparatus consists of two Columbia dry cells No. 6, and a Weston D.C. voltmeter connected in series. The voltmeter had a resistance of 251 ohms, and read 150 with the circuit closed. As would naturally be expected the needle points introduce additional resistance; and differences in readings arise from varying pressure with which the points are pressed upon the mineral surface, the fact that soft minerals allow the points to penetrate further tnan do hard minerals; and that a rough surface will give a different reading from the same surface when perfectly polished. This method is, therefore, qualitative rather than quantitative, but does serve to arrange the minerals in a series in proper order, beginning with those that do not conduct electricity, and ending with those that conduct electricity as easily as copper. Only minerals that are found widely apart in this series may be definitely distinguished by this means. The mineral to be tested is brought into the center of the field of the microscope, the two steel needle points are placed upon the surface, and the voltmeter reading noted. The mineral will immediately give either a reading of zero, of 150, or an intermediate reading; thus, giving at once three groups: minerals apparently non-conductors, minerals with conductivity equal to or greater than copper, and minerals having electrical conductivity but with greater resistance than copper. Minerals belonging to these three groups will be found tabulated in Table 10, page 123. By this method plagionite may be distinguished from galena, stromeyerite from argentite, tennantite from tetrahedrite, pentlandite from chalmersite, etc.

Electrolysis may be used to etch and bring out the structure of minerals in polished sections. After a drop of the reagent is placed on the polished surface under the microscope, it was found that the chemical action could be "speeded up" by the use of a weak electric current. The current used was furnished by one dry cell of the same type as used in determining the comparative electrical conductivity; and the terminals were a sewing needle and a piece of fine platinum wire; the former being connected to the carbon pole, and the latter to the zinc pole of the dry battery. The needle is placed on the polished surface just outside of the drop of reagent, and the platinum wire brought into contact with the top of the drop of reagent. Using a 20 per cent. solution of $FeCl_3$ a clot of metallic copper will be produced on the end of the platinum wire by algodonite, bornite, chalcocite, native copper, and covelite. With the same reagent a beautiful "wood grain" structure is developed on enargite, famatinite, and luzonite, which distinguishes them from tetrahedrite and tennantite. The effects of a weak current on the microchemical reactions may be generalized as follows: (1) a reagent that is negative when used alone may be strongly reactive with the current; (2) any reagent that reacts alone will give a much more intense reaction with the current; and (3) if the reagent with the current gives no reaction, it is a good check that the reagent is truly negative.

The electro-potentials of minerals have been investigated not only as a possible means of identifying minerals, but also because of their important bearing upon the processes of enrichment of ore deposits. Although the results obtained cannot be considered as of general value for the purpose of identification, still a summary of the work thus far accomplished is suggestive and may well lead to further investigation. The method used in making the test is very simple. A small drop of dilute nitric acid is applied to the mineral while under the microscope. The polished section is then removed and a minute primary cell is established by bringing a standard electrode into contact with the top of the drop and completing the circuit, into which has been introduced a millivoltmeter, with a wire placed on the mineral surface near the drop. Eight standard electrodes, which are wires composed of a solid solution of copper and gold and containing, respectively, 100 per cent., 95 per cent., 90 per cent., 80 per cent., 70 per cent., 60 per cent., 50 per cent. and 40 per cent. of copper, are used to determine the direction and intensity of the electric potential difference. The results of testing many of the minerals having high electrical conductivity are tabulated in Table 11, page 124, and indicate the position of the more common ore minerals in the electromotive series. These tests have emphasized the fact that when an active reagent is placed upon the contact of two electrically conductive minerals, the reagent becomes the electrolyte of a primary cell in which the minerals in contact are electrodes, and a weak current will be produced. The result is that the mineral of lower potential will dissolve more rapidly than normally, and this demonstrates the fact that microchemical tests should only be made upon pure material, and that the influence of impurities may give conflicting microchemical observations.

The following reagents have been used in this work and their effects are briefly described in the determinative tables. They are of the same strength as those used by Murdoch and thus it has been possible to compare directly the results obtained.

HNO_3	One part concentrated acid (sp. gr. = 1.42) and one part water.
HCl	One part concentrated acid (sp. gr. = 1.19) and one part water.
KCN	20 per cent. solution in water.
$FeCl_3$	20 per cent. solution in water.
$HgCl_2$	Saturated solution in water.
KOH	Saturated solution in water.

Care should be taken to use reagents of approximately the above strength as discordant results are otherwise often obtained. Care should be taken to confine the reagent to the surface of the mineral under examination also because reactions with soluble gangue substances sometimes confuse the results.

Finally, qualitative chemical tests for the elements can be applied under the microscope.[1] They are of great value in certain special problems and the mineragrapher must more and more resort to them. As a rule two or sometimes three reagents only are required for such a test and, with very little practice, results of such nicety can be obtained that, once used, microchemistry will always hold its place. A condensed list of the microchemical tests for elements commonly found in the ore minerals is given in Table 12, page 145.

[1] See CHAMOT, E. M. "Elementary Chemical Microscopy." New York, 1915.

CHAPTER II

PHOTOMICROGRAPHY OF POLISHED SECTIONS

For the mining geologist, photographs of the ore minerals and the relations between them, as seen by vertical illumination are almost indispensable in preparing reports. A well-chosen illustration will establish a point more conclusively in the minds of the readers than pages of written description. For the student of ore deposits, such microphotographs are invaluable records to be preserved as reference material.

The arrangement is simple, there being a camera in which a microscope takes the place of the lens. Practically, however, good results require a high-grade specially designed microscope and strong artificial light. The procedure has been borrowed largely from metallography, except that metallographers seldom have to bring out color values by optical means, while in mineragraphy the variety of different minerals in one section calls for rather refined methods of color screening or filtering. The large Leitz metallographic microscope has been used by the writers, but any good instrument made on the same optical principles such as the Leitz type manufactured by Bausch and Lomb would serve equally well. A 5-ampere D.C. arc lamp is used as the source of light with this microscope. A description of the construction and specialized directions for the use of such apparatus are out of place here and are always supplied by the makers with the instruments. We are, however, concerned with many factors dealing with the choice of subject and the technique of photomicrography in general.

First, it is well worth the time to search carefully over all the polished sections available in order to locate an area which will best illustrate as many as possible of the important features in one picture. In too many microphotographs which have appeared in the literature the reader has to accept on faith the writer's statement that some relationship is established by the illustration. It is true that two people may not always interpret a particular structure in the same way, but it is a rare specimen which justifies positive conclusions and which does not show an area illustrating them.

When the subject for the photograph has been chosen care should be taken to adjust the reflector in such a way as to obtain uniform illumination over the entire field, otherwise a negative of uneven density will result.

The proper color screen to bring out the contrasts between the various minerals must now be selected. The sensitiveness of the photographic plate to light of different colors or wave lengths is very different from that of the human eye. Ordinary white light and light from most artificial sources is a mixture of lights of all wave lengths. The eye sees all of these colors through a rather wide range, while the photographic plate "sees" colors outside of the limits of this range and is "blind" to much that is within it. For example the eye registers red as a bright color, but the ordinary plate is unaffected by red light and in a picture all red objects would appear black. For a complete and interesting treatment of this subject the reader is referred to "The Photography of Colored Objects" by Dr. C. E. Kenneth Mees.[1] Certain dyes have the property of absorbing light of certain wave lengths and by passing the light to be used through filters made with these dyes, color effects can be controlled upon the negative to a large extent. Table 1 lists the Wratten M filters supplied by the Eastman Kodak Company and indicates the approximate range of light transmitted by each.

Experience has shown that a filter should be used at all times as several optical difficulties arise when photography is attempted without them especially at magnifications greater than 100 diameters. For the average section with a variety of minerals clearly distinguishable by the eye the K_3 yellow filter most nearly reproduces the true values as seen (that is when used with the Wratten M plate described below). In many cases, however, when strong contrast is required between two minerals it is necessary to select one of the special filters described in Table 1, or, sometimes, one of the combinations of two filters. This is done usually by trial, the contrast obtained with each mineral being observed upon the ground glass screen. This visual trial method is only trustworthy when using a panchromatic plate which approximates the relative sensitiveness of the eye. The best panchromatic plate obtainable for this type of work is the Wratten M plate distributed by the East-

[1] Mees, C. E. Kenneth. "The Photography of Coloured Objects." *Wratten and Wainwright, Ltd., London,* 1913.

TABLE 1[1]

Wratten filter	Visual color	Range of spectral transmission in $\mu\mu$ of wave length
A	Scarlet	From 580 to red end
B	Green	From 460 to 600
C	Blue-violet	From 400 to 510
D	Purple	From 380 to 460 and from 640 to red end
E	Orange	From 560 to red end
F	Pure red	From 610 to red end
G	Strong yellow	From 510 to red end
H	Blue	From 420 to 540
K_1	Pale yellow (very pale)	Wide range of lower spectrum
K_2	Pale yellow	
K_3	Yellow	
D and H	Violet	About 450
C and H	Blue	About 480
B and C	Blue-green	About 500
B and H	Bluish-green	About 520
G and H	Pure green	About 535
B and G	Yellowish-green	About 550
B and E	Greenish-yellow	About 575
A and D	Deep red	About 660

man Kodak Company. It has all the color sensitiveness of the true panchromatic plate, being even more sensitive to red light, and has an extremely fine-grained emulsion especially designed for microscopic photography. The worker along this line cannot do better than to use it exclusively. All of the data concerning proper filters and some of the factors for calculating exposures contained in this chapter are applicable only for use with the Wratten M plate.

Having chosen the proper filter, accurate focusing is the next important step. It is impossible to obtain sharp, clear cut negatives by focusing upon the ground glass screen alone. After an approximate focus is obtained on this screen it should be replaced by a plain glass screen upon which the final focus is made with the use of a lens. This screen should be very carefully removed and the photographic plate put in its place. Everything should now be ready for the exposure.

The data for calculating the proper exposure for different magnifications, sources of light, filters, etc., has been assembled

[1] *From the booklet "Photomicrography" published by the Eastman Kodak Co.*

by the Eastman Kodak Company and Tables 2 and 5 have been taken from their booklets.[1] The formula for calculating exposures gives an approximate figure which should be tested experimentally for different types of subjects. The formula does not take this latter factor into consideration, but the same exposure obviously will not give equally good results on a section composed largely of white minerals and one in which dark gray minerals are dominant. The method of making the test exposure is as follows: If the formula indicates that an exposure of 20 sec. is about right, expose successive portions of the plate for 5, 10, 20, 40, and 80 sec. by pulling out the slide and then pushing it in one-fifth the width of the plate at the end of each of the above periods. The exposure should increase by geometric progression as indicated rather than by arithmetical progression such as 10, 15, 20, 25, and 30 sec., as the difference in density of negatives exposed 10 and 20 sec. respectively is twice as great as in those exposed 20 and 30 sec. respectively. With this test plate as a guide, the proper exposure can be easily determined.

In making the original exposure calculation the following formula is used:

Standard exposure × N.A. factor × source factor × magnification factor × filter factor = exposure in seconds.

The standard exposure used in the above empirical equation is taken as 10 sec.

The exposure varies inversely as the numerical aperture, N.A. of the objective. (The numerical aperture is an optical constant for a lens of given focal length.) The factors for the N.A. of the objectives are given in Table 2.

The source of light naturally is of the greatest importance in determining the length of exposure. In past publications the factors for the source of light have been adopted from data for transmitted light conditions. This has led to confusion because the polished surface does not reflect all of the light by any means, and consequently requires a longer exposure. Experience has shown that, on the average, the factor for reflected light is two to four times that for direct transmitted illumination. These factors for various sources of light are given in Table 3.

[1] *"Photomicrography, Color Plates and Filters for Commercial Photography, and Reproduction Work with Dry Plates." Eastman Kodak Company, Rochester, N. Y.*

TABLE 2

Focal length objective		Average N.A.	Approximate exposure factor
Inches	Millimeters		
4	100	0.08	40
3	75	0.09	30
2	50	0.15	10
1	25	0.25	4
2/3	16	0.35	2
1/2	12	0.45	1 1/4
1/3	8	0.50	1
1/4	6	0.80	2/5
1/6	4	0.85	1/3
1/8	3	0.90	1/3
1/12	2	1.30	1/4

TABLE 3

Source of light	Exposure factor for M plate
Oil	3
Incandescent gas	1
Nernst lamp (1 amp.)	1/4
Acetylene	1/4
Direct current arc (5 amps.)	1/8
Direct current arc (10 amps.)	1/33
Direct current arc (30 amps.)	1/330

(It should be understood that the above factors are by nature very approximate, dependent entirely on arrangement and power.)

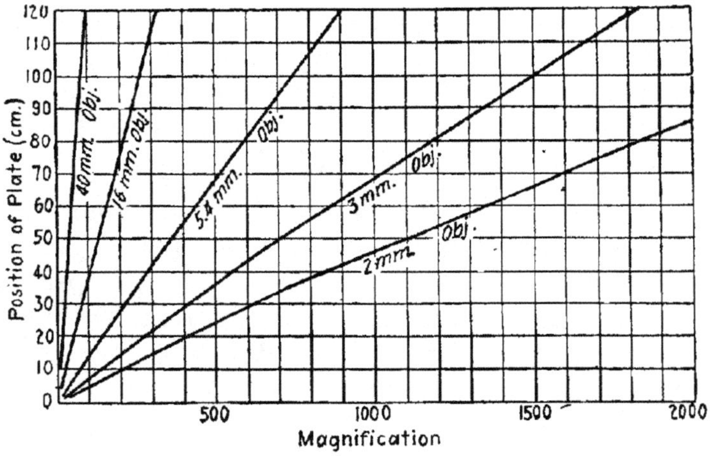

FIG. 3.

The greater the magnification, the longer is the required exposure. The degree of magnification depends upon the focal length of the objective, the projection eyepiece, the distance of the photographic plate from the objective, etc. It will be found convenient to prepare a set of curves for the instrument used, similar to those in Fig. 3, which have been calculated for the Leitz metallographic microscope.

In Table 4 the exposure factors for different magnifications are listed.

TABLE 4

Magnification	Exposure factor for M plate	Magnification	Exposure factor for M plate
10	1/100	600	36
25	1/16	750	56
50	1/4	900	81
100	1	1000	100
250	6	2000	400
500	25	2500	625

The factor for any magnification may be calculated on the basis that it increases with the square of the diameter.

The last variable in the exposure equation is the factor for the filter used. These factors are dependent upon the source of illumination since the different artificial lights have widely differing color values. Table 5 lists the multiplying factors for the filters, singly and in pairs, calculated for the open arc and the Wratten M plate.

TABLE 5

Filter	Factor (M plate and open arc)	Filter	Factor (M plate and open arc)
A	10	K_2	3
B	18	K_3	4½
C	12	D and G	250
D	Only used in combination	A and D	240
E	8	B and E	250
F	15	G and H	1600
G	8	B and C	600
H	15	B and G	60
K_1	1½	D and H	60

The foregoing tables are used as follows: the approximate exposure for a photograph using a 16-mm. objective (Leitz

No. 3), a 5-amp. D.C. arc, magnification 50, and screen K_3 would be:

$$10 \text{ sec.} \times 2 \times 1/8 \times 1/4 \times 9/2 = 3 \text{ sec. (approximately)}$$

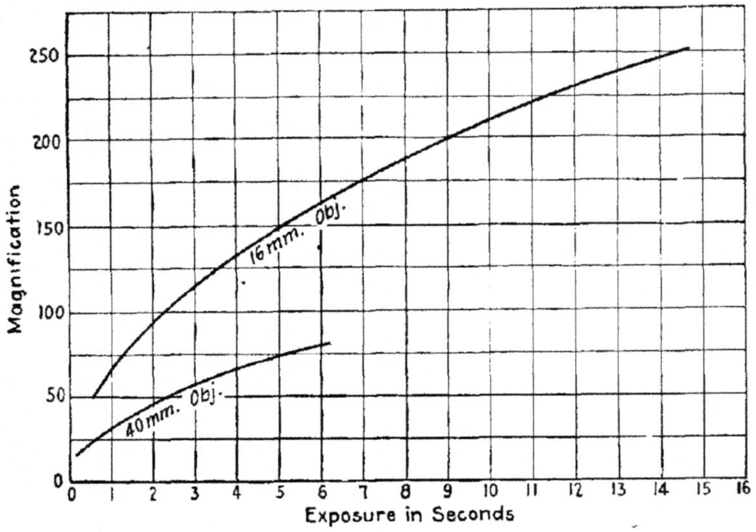

FIG. 4.

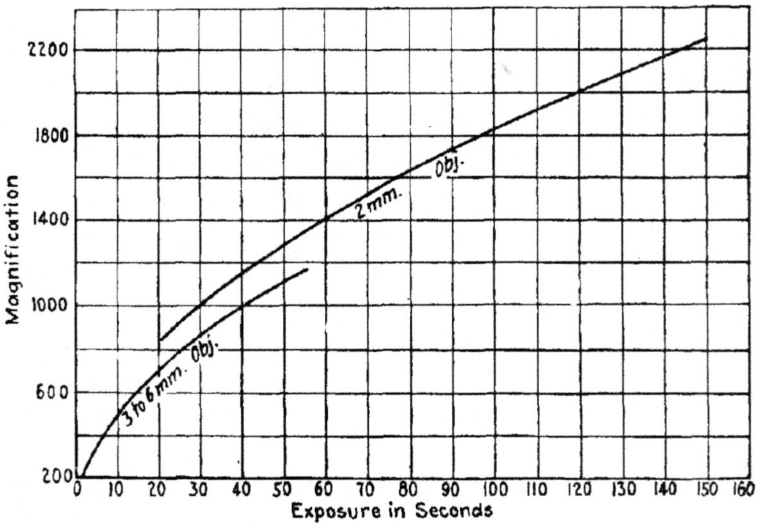

FIG. 5.

Figures 4 and 5 give the approximate exposures for an average specimen consisting, say, of galena, chalcopyrite, and sphalerite. These curves are calculated solely for the Wratten M plate. They are for the 5 amp. open arc, wide open stop, and without filter. When a filter is used the value obtained from the curves should be multiplied directly by the filter factor taken from Table 5. These curves are plotted from theoretical values checked in practice and may be multiplied by a suitable factor when another light source is used.

The Wratten plate is very sensitive to red light and consequently must be opened and developed either in total darkness or in the light of the Wratten Safelight, series 3, a faint *green* light. With each box of plates an instruction sheet is enclosed giving the developing formula, proper temperatures, and times of development. These instructions, when followed, yield excellent results and should only be varied by the most experienced worker. In the course of the dark room treatment it will be found advisable to wash the carbon backing off the plates with a soft sponge while rinsing before placing in the fixing solution.

In preparing positives a glossy paper usually yields the most satisfactory detail and reproduces best. After developing and fixing the print should be pressed face down upon a ferrotype plate with a rubber roller and allowed to dry thoroughly, when they will peel off without sticking providing a sufficient amount of hardening solution has been used in the fixing bath. Small amounts of a special polishing solution can be used on the ferrotype plate when necessary and will eliminate sticking entirely. Such a solution is sold by Burke and James of Chicago, Ill.

CHAPTER III

THE USE OF THE DETERMINATIVE TABLES

The tables have been arranged with the idea of enabling the student to apply a uniform series of simple and positive tests at one time to all the different minerals present in appreciable amounts in the section. The following procedure in making determinations has been found to yield rapid and exact results. It must be appreciated that, as in all other things, experience and practice make for greater skill and, although it has been demonstrated that beginners can mechanically follow the tables with fair success, familiarity with the appearance and microchemical behavior of the common ore minerals naturally gives confidence to the worker. No one should attempt a serious application of this method of study to any problem without first applying it to a few of the common minerals.

Fig. 6.

The polished section of ore pictured in Fig. 6 has been chosen as an example. This particular ore is of value in this connection because it illustrates the variety of tests which go toward exact mineral determinations under the metallographic microscope.

The preliminary examination of the material should be done

systematically and the results recorded somewhat as in Table 6. The different minerals should be temporarily designated by number and a brief description of their appearance. The reagents used in the scheme of identification are preferably applied in their order in the tables and the reactions, if any, observed and recorded.

TABLE 6

	1. Creamy white mineral with high relief	2. Yellow	3. Violet white
HNO_3	Unaffected for some time, but slowly tarnishes brown with very slow effervescence	Neg.	Slowly effervesces and turns bluish
HCl	Neg.	Neg.	Neg.
KCN	Neg.	Neg.	Neg.
Hardness	Very high	Medium	High to medium
$FeCl_3$	Neg.	Neg.	Neg.
	(Pyrite)	(Chalcopyrite)	(Polydymite)

In first running down the unknown it is often most convenient to use the outline of the determinative tables, especially if the worker is familiar with the more common minerals by sight, because larger groups of minerals are seen at one time and slight uncertainties as to the results of one or more tests can be eliminated. (1) Readily falls out as pyrite, (2) as chalcopyrite, and (3) is limited to a small group of minerals. Of these, arsenopyrite, willyamite, kallilite, and niccolite all contain arsenic or antimony. Blowpipe tests applied to a small piece of the material gouged from the edge reveal neither of these elements and polydymite is indicated by a process of elimination.

However, since the mineral was dissolved readily by nitric acid, a qualitative microchemical test for nickel should be easily applied. (See Table 12, page 145, for list of these tests.) After a drop of nitric acid has remained on the supposed polydymite for a minute or two a drop of tartaric acid is added to keep iron in solution when a drop of concentrated ammonium hydroxide is next added to make the solution alkaline. If a drop of a solution of dimethyl glyoxime in alcohol is now added, a beautiful scarlet-red crystalline precipitate appears under the microscope. This proves the presence of nickel in the unknown and leads to the conclusion that it is polydymite.

The foregoing illustrates the general method to be followed in making identity determinations. All reactions and properties are utilized when necessary and the individual must be the judge as to how much evidence is needed in each case before reaching a decision. This in turn is dependent upon the individual's experience and familiarity with the properties of the different minerals as seen under the microscope in reflected light.

Abbreviations used in the Tables

B.B.	=	Before the blowpipe.
C.	=	Color seen megascopically.
Fus.	=	Fusibility.
HNO_3 -E	=	Reacts with HNO_3 with effervescence.
HNO_3	=	Visible reaction such as tarnish, etc. (Without effervescence in the case of HNO_3.)
HNO_3 -N	=	No visible reaction with HNO_3.
H	=	Hardness high.
L	=	Hardness low.
M	=	Hardness medium.
Microchem.	=	Microchemical tests.
O.F.	=	Oxidizing flame.
R.F.	=	Reducing flame.
S.Ph.	=	Salt of Phosphorus.
Str.	=	Streak.

OUTLINE OF THE DETERMINATIVE TABLE

Effervesces with HNO_3

				Color	Mineral	Formula	Page
HCl	KCN	M	$FeCl_3$	Cream	*Whitneyite*	Cu_9As	33
				Creamy white	*Huntilite*	Ag_3As	33
				Grayish white	**Cuprite**	Cu_2O	33
		L	$FeCl_3$	Creamy white	*Chilenite*	Ag_6Bi	35
				Creamy white	*Domeykite*	Cu_3As	35
				Purple	*Rickardite*	Cu_4Te_3	35
				Pink	**(Native copper)**[1]	Cu	51
				Bluish white	**(Chalcocite)**	Cu_2S	51
				Galena white	(HESSITE)	Ag_2Te	66
			$FeCl_3$–N	Creamy white	(*Aikinite*)	$3(Cu_2Pb)S.Bi_2S_3$	63
	KCN–N	H	$FeCl_3$–N	White (violet)	(*Polydymite*)	Ni_4S_5	57
		M	$FeCl_3$–N	Grayish white	ALABANDITE	MnS	41
		L	$FeCl_3$	White	*Naumannite*	$(Ag_2Pb)Se$	43
				White	*Altaite*	PbTe	43
				Creamy white	**Native silver**	Ag	43
				Creamy white	*Native bismuth*	Bi	43
				Creamy white	*Tapalpaite*	$3Ag_2(STe).Bi_2(STe)_3$	43
				Galena white	*Plagionite*	$5PbS.4Sb_2S_3$	43
				Galena white	**(Galena)**	PbS	77
			$FeCl_3$–N	Galena white	*Semseyite*	$7PbS.3Sb_2S_3$	45
				White	(*Aikinite*)	$3(Cu_2Pb)S.Bi_2S_3$	63
				Galena white	(*Plagionite*)	$5PbS.4Sb_2S_3$	43
				Galena white	(*Dognacskaite*)	Cu, Bi, and S	63
				Grayish white	(JAMESONITE)	$2PbS.Sb_2S_3$	62
				Grayish white	(ALABANDITE)	MnS	41
				Galena white	(*Horsfordite*)	Cu_6Sb	63
HCl–N	KCN	H	$FeCl_3$–N	White	(*Polydymite*)	Ni_4S_5	57
		M	$FeCl_3$	Pink	**(Native copper)**	Cu	51
		L	$FeCl_3$	Pink	**Native copper**	Cu	51
				Bluish white	**Chalcocite**	Cu_2S	51
				Pinkish brown	**(Bornite)**	Cu_5FeS_4	53
			$FeCl_3$–N	Pinkish brown	**Bornite**	Cu_5FeS_4	53
				Creamy white	(*Aikinite*)	$3(PbCu_2)S.Bi_2S_3$	63
				Galena white	**(Stibnite)**	Sb_2S_3	87
	KCN–N	H	$FeCl_3$	Creamy pink	**Niccolite**	NiAs	55
				Pinkish white	*Maucherite*	Ni_3As_2	55
				White	**Smaltite**	$CoAs_2$	55

[1] Parentheses about a mineral indicate that it is not in its normal position.

Effervesces with HNO_3

			Color	Mineral	Formula	Page
		$FeCl_3$–N	White	**Arsenopyrite**	FeAsS	57
			Violet white	*Polydymite*	Ni_4S_5	57
			White	*Willyamite*	(CoNi)SbS	57
			White	*Kallilite*	Ni(SbBi)S	57
			Creamy white	**Pyrite**	FeS_2	57
			Creamy white	**Marcasite**	FeS_2	57
			Creamy pink	(**Niccolite**)	NiAs	55
	M	$FeCl_3$	Creamy white	*Cosalite*	$Pb_2Bi_2S_5$	59
			Galena white	*Native arsenic*	As	59
	L	$FeCl_3$	Cream	*Melonite*	Ni_2Te_3	61
			Creamy white	CALAVERITE	$AuTe_2$–Ag	61
			White	*Native tellurium*	Te	61
			Galena white	*Rezbanyite*	$4PbS.5Bi_2S_3$	61
			Galena white	*Tetradymite*	$Bi_2(Te, S)_3$	61
			Creamy white	(KRENNERITE)	$AuTe_2$–Ag	63
			Creamy white	(**Native silver**)	Ag	43
			Galana white	(*Plagionite*)	$5PbS.4Sb_2S_3$	43
			Grayish white	(*Petzite*)	$(Ag, Au)_2Te$	97
		$FeCl_3$–N	Creamy white	KRENNERITE	$AuTe_2$–Ag	63
			Creamy white	*Emplectite*	$CuBiS_2$	63
			Creamy white	*Chiviatite*	$Pb_2Bi_2S_{11}$	63
			Creamy white	*Aikinite*	$3(PbCu_2)S\ Bi_2S_3$	63
			Galena white	*Dognacskaite*	Cu, Bi, and S	63
			Galena white	*Boulangerite*	$3PbS.Sb_2S_3$	63
			Galena white	*Horsfordite*	Cu_6Sb	63
			Galena white	*Bismuthinite*	Bi_2S_3	62
			Grayish white	REALGAR	AsS	62
			Grayish white	JAMESONITE	$2PbS.Sb_2S_3$	62
			Grayish white	*Zinkenite*	$PbS.Sb_2S_3$	62
			Bluish Gray	*Tungstenite*	WS_2	62
			White	(*Native tellurium*)	Te	61
			Galena white	(**Stibnite**)	Sb_2S_3	87

Reacts with HNO_3—does not Effervesce

				Color	Mineral	Formula	Page
HCl	KCN	M	$FeCl_3$–N	Creamy white	*Sulvanite*	Cu_3VS_4	65
				Galena white	(*Guejarite*)	$Cu_2Sb_4S_7$	83
		L	$FeCl_3$	Creamy white	*Dyscrasite*	Ag_3Sb, Ag_6Sb, etc.	67
				Galena white	Hessite	Ag_2Te	66
				Grayish white	**Argentite**	Ag_2S	67
				Grayish white	*Stromeyerite*	$(Ag,Cu)_2S$	67
				Grayish white	*Jalpaite*	$3Ag_2S.Cu_2S$	67
				Brown	(**Pyrolusite**)	MnO_2	101
			$FeCl_3$–N	Bluish white	*Brongniardite*	$PbS.Ag_2S.Sb_2S_3$	69
	KCN–N	H	$FeCl_3$–N	Gray	PSILOMELANE	H_4MnO_6	71
		M	$FeCl_3$	Grayish white	TENORITE	CuO	73
			$FeCl_3$–N	Gray	*Cuprodescloizite*	$(PbZnCu)_4V_2O_9.H_2O$	75
				Gray	(**Sphalerite**)	ZnS	94
				Grayish white	(TENORITE)	CuO	73
		L	$FeCl_3$	Purple	*Umangite*	Cu_3Se_2	77
				White	*Clausthalite*	$PbSe$	77
				Galena white	**Galena**	PbS	77
				Galena white	*Lillianite*	$3PbS.Bi_2S_3$	77
				Grayish white	*Teallite*	$PbSnS_2$	77
				Grayish white	*Cylindrite*	$Pb_6Sb_2Sn_6S_{21}$	77
				Grayish white	*Aguilarite*	$Ag_2S.Ag_2Se$	76
				White	(*Native antimony*)	Sb	96
				Creamy white	(**Native silver**)	Ag	43
			$FeCl_3$–N	Galena white	*Meneghinite*	$4PbS.Sb_2S_3$	79
				Galena white	*Geocronite*	$5PbS.Sb_2S_3$	79
				Grayish white	*Franckeite*	$Pb_5Sn_3Sb_2S_{12}$	79
				White	(*Eucairite*)	$Cu_2Se.Ag_2Se$	97
				Grayish white	(*Cylindrite*)	$Pb_6Sb_2Sn_6S_{21}$	77
HCl–N	KCN	M	$FeCl_3$	Grayish white	(POLYBASITE)	Ag_9SbS_6	87
			$FeCl_3$–N	Pinkish white	*Luzonite*	Cu_3AsS_4	83
				Brownish white	**Enargite**	Cu_3AsS_4	83
				Pinkish white	FAMATINITE	Cu_3SbS_4	83
				Galena white	*Guejarite*	$Cu_2Sb_4S_7$	83
				Grayish white	*Freibergite*	$(Cu, Ag)_8Sb_2S_7$	82
				Yellow	(**Chalcopyrite**)	$CuFeS_2$	117
				Creamy white	(*Sulvanite*)	Cu_3VS_4	65
				Grayish white	(**Tennantite**)	$Cu_8As_2S_7$	117
				Grayish white	(**Tetrahedrite**)	$Cu_8Sb_2S_7$	117
		L	$FeCl_3$	Grayish white	*Pearceite*	Ag_9AsS_6	85
				Grayish white	(**Argentite**)	Ag_2S	67
				Grayish white	(POLYBASITE)	Ag_9SbS_6	87
				Grayish white	(*Jalpaite*)	$3Ag_2S,Cu_2S$	67
				Bluish white	(*Polyargyrite*)	$Ag_{24}Sb_2S_{15}$	109

Reacts with HNO_3—does not Effervesce

			Color	Mineral	Formula	Page
KCN-N	H	$FeCl_3$-N	Galena white	**Stibnite**	Sb_2S_3	87
			Galena white	KERMESITE	Sb_2S_2O	87
			Grayish white	POLYBASITE	Ag_9SbS_6	87
			Galena white	(*Livingstonite*)	$HgSb_4S_7$	111
		$FeCl_3$	White	**Chloanthite**	$NiAs_2$	89
			White	*Rammelsbergite*	$NiAs_2$	89
			White	GERSDORFFITE	NiAsS	89
		$FeCl_3$-N	White	*Löllingite*	$FeAs_2$	91
			White	*Ullmannite*	NiSbS	91
			Creamy white	*Linnæite*	Co_3S_4	91
			Pinkish white	GLAUCODOT	(Co,Fe)AsS	91
			Cream	*Hauchecornite*	$(NiCo)_7(SSbBi)_8$	91
			Cream	(**Pyrrhotite**)	$FeS(S)_x$	95
			Creamy white	(**Pyrite**)	FeS_2	57
			White	(**Chloanthite**)	$NiAs_2$	89
			White	(GERSDORFFITE)	NiAsS	89
			White	(*Willyamite*)	(CoNi)SbS	57
	M	$FeCl_3$	Pink	*Breithauptite*	NiSb	93
		$FeCl_3$-N	Pale yellow	MILLERITE	NiS	95
			Cream	**Pyrrhotite**	$FeS(S)_x$	95
			Cream	*Chalmersite*	$CuFe_2S_3$	95
			Creamy white	PENTLANDITE	(Fe,Ni)S	95
			Creamy white	*Wittichenite*	Cu_3BiS_3	95
			Galena white	*Andorite*	$PbAgSb_3S_6$	95
			Grayish white	BOURNONITE	$(PbCu_2)_3Sb_2S_6$	94
			Grayish white	*Stylotypite*	$3(Cu_2Ag_2Fe)$- $S.Sb_2S_3$	94
			Grayish white	*Stannite*	$SnCu_2FeS_4$	94
			Grayish white	*Crookesite*	$(CuTlAg)_2Se$	94
			Grayish white	*Hauerite*	MnS_2	94
			Gray	**Sphalerite**	ZnS	94
			Yellow	(**Chalcopyrite**)	$CuFeS_2$	117
			Creamy white	(*Sulvanite*)	Cu_3VS_4	65
			Pinkish white	(FAMATINITE)	Cu_3SbS_4	83
			Gray	(*Voltzite*)	Zn_5S_4O	116
			Grayish white	(**Tennantite**)	$Cu_8As_2S_7$	117
	L	$FeCl_3$	White	*Berzelianite*	Cu_2Se	97
			White	*Eucairite*	$Cu_2Se.Ag_2Se$	97
			White	*Native antimony*	Sb	96
			Creamy white	*Kalgoorlite*	$HgAu_2Ag_6Te_6$	97
			Grayish white	*Petzite*	$(Ag,Au)_2Te$	97
			Grayish white	*Coloradoite*	HgTe	97
			Grayish white	**Molybdenite**	MoS_2	97
			Creamy white	(**Native silver**)	Ag	43
			Galena white	(*Plagionite*)	$5PbS.4Sb_2S_3$	43
			Grayish white	(*Aguilarite*)	$Ag_2S.Ag_2Se$	76

Reacts with HNO_3—does not Effervesce

	Color	Mineral	Formula	Page
$FeCl_3$–N	Pale brown	*Sternbergite*	$AgFe_2S_3$	99
	Galena white	*Lengenbachite*	$Pb_6(AgCu)_2As_4S_{13}$	99
	Galena white	*Rathite*	Pb(AsSb)S	99
	Galena white	*Jordanite*	$Pb_4As_2S_7$	99
	Galena white	*Guitermanite*	$3PbS.As_2S_3$	99
	Galena white	*Epiboulangerite*	$Pb_3Sb_2S_8$	99
	Galena white	*Galenobismutite*	$PbBi_2S_4$	98
	Galena white	*Beegerite*	$Pb_6Bi_2S_9$	98
	Galena white	*Freieslebenite*	$(PbAg_2)_5Sb_4S_{11}$	98
	Galena white	*Nagyagite*	$Au_2Pb_{10}Sb_2Te_6S_{15}$	98
	Creamy white	*Sylvanite*	$AuAgTe_2$	98
	Creamy white	*Guanajuatite*	Bi_2Se_3	98
	White	(*Eucairite*)	$Cu_2Se.Ag_2Se$	97
	White	(*Berzelianite*)	Cu_2Se	97
	Creamy white	(*Emplectite*)	$CuBiS_2$	63
	Galena white	(**Stibnite**)	Sb_2S_3	87
	Galena white	(*Geocronite*)	$5PbS.Sb_2S_3$	79
	Galena white	(*Andorite*)	$PbAgSb_3S_6$	95
	Grayish white	(*Crookesite*)	$(CuTlAg)_2Se$	94
	Grayish white	(*Regnolite*)	$Cu_7As_2S_{12}$	119
	Grayish white	(JAMESONITE)	$2PbS.Sb_2S_3$	62

Does not React with HNO_3

				Color	Mineral	Formula	Page
HCl	KCN	L	$FeCl_3$	Grayish white	STEPHANITE	Ag_5SbS_4	101
				Brown	**Pyrolusite**	MnO_2	101
				Grayish white	(*Jalpaite*)	$3Ag_2S.Cu_2S$	67
				Bluish white	(**Proustite**)	Ag_3AsS_3	109
	KCN–N	M	$FeCl_3$–N	Grayish white	*Delafossite*	CuO, Fe_2O_3	103
HCl–N	KCN	M	$FeCl_3$	Grayish white	(POLYBASITE)	Ag_9SbS_6	87
			$FeCl_3$–N	Yellow	(**Chalcopyrite**)	$CuFeS_2$	117
				Brownish white	(**Enargite**)	Cu_3AsS_4	83
				Grayish white	(PYRARGYRITE)	Ag_3SbS_3	111
		L	$FeCl_3$	Bluish white	**Proustite**	Ag_3AsS_3	109
				Bluish white	*Polyargyrite*	$Ag_{24}Sb_2S_{15}$	109
				Grayish white	**Cerargyrite**	AgCl	109
				Grayish white	(*Jalpaite*)	$3Ag_2S$ Cu_2S	67
				Grayish white	(*Pearcite*)	Ag_9AsS_6	85
			$FeCl_3$–N	Blue	**Covellite**	CuS	111
				Yellow	**Native gold**	Au	111
				Galena white	*Livingstonite*	$HgSb_4S_7$	111
				Bluish white	*Onofrite*	Hg(SSe)	111
				Grayish white	PYRARGYRITE	Ag_3SbS_3	111
				Bluish white	*Miargyrite*	$AgSbS_2$	110
				Grayish white	*Argyrodite*	Ag_6GeS_5	110
				Grayish white	ORPIMENT	As_2S_3	110
				Bluish white	(**Proustite**)	Ag_3AsS_3	109
				Grayish white	(POLYBASITE)	Ag_9SbS_6	87
	KCN–N	H	$FeCl_3$	Gray	*Uraninite*	UO_3, etc.	113
			$FeCl_3$–N	White	*Sperrylite*	$PtAs_2$	115
				Pinkish white	**Cobaltite**	CoAsS	115
				Creamy white	**Hematite**	Fe_2O_3	115
				Grayish white	**Magnetite**	Fe_3O_4	115
				Grayish white	**Ilmenite**	$FeTiO_3$	115
				Grayish white	**Rutile**	TiO_2	115
				Grayish white	*Franklinite*	(FeZnMn)O. $(FeMn)_2O_3$	115
				Grayish white	MANGANESE OXIDES		114
				Grayish white	WOLFRAMITE	$(FeMn)WO_4$	114
				Grayish white	*Rare earths*		114
				Gray	CASSITERITE	SnO_2	114
				Gray (variable)	**Limonite**	$2Fe_2O_3.3H_2O$	114
				Gray	CHROMITE	$FeCr_2O_4$	114
				White	(*Löllingite*)	$FeAs_2$	91

Does not React with HNO_3

		Color	Mineral	Formula	Page
M	$FeCl_3$–N	Yellow	**Chalcopyrite**	$CuFeS_2$	117
		Galena white	*Chalcostibite*	$CuSbS_2$	117
		Galena white	*Berthierite*	$FeSb_2S_4$	117
		Grayish white	*Baumhauerite*	$Pb_4As_6S_{13}$	117
		Grayish white	**Tetrahedrite**	$Cu_8Sb_2S_7$	117
		Grayish white	**Tennantite**	$Cu_8As_2S_7$	117
		Gray	*Voltzite*	Zn_5S_4O	116
		Gray	*Erythrozincite*	$(MnZn)S$	116
		Pale yellow	(MILLERITE)	NiS	95
		Cream	**(Pyrrhotite)**	$FeS(S)_x$	95
		Cream	(*Chalmersite*)	$CuFe_2S_3$	95
		Creamy white	(PENTLANDITE)	$(Fe,Ni)S$	95
		Grayish white	(BOURNONITE)	$(PbCu_2)_3Sb_2S_6$	94
L	$FeCl_3$–N	Galena white	*Dufrenoysite*	$Pb_2As_2S_5$	119
		Galena white	*Matildite*	$AgBiS_2$	119
		Galena white	*Lehrbachite*	$PbSe + HgSe$	119
		Bluish white	*Tiemannite*	$HgSe$	119
		Bluish white	*Vrbaite*	$TlAs_2SbS_5$	119
		Bluish white	*Stützite*	Ag_4Te	119
		Grayish white	*Regnolite*	$Cu_7As_2S_{12}$	119
		Grayish white	*Seligmannite*	$CuPbAsS_3$	119
		Grayish white	**Cinnabar**	HgS	118
		Grayish white	*Metacinnabarite*	HgS	118
		Grayish white	*Patronite*	VS_4	118
		Gray	*Lorandite*	$TlAsS_2$	118
		Galena white	(*Freieslebentie*)	$(Pb, Ag_2)_5Sb_4S_{11}$	98
		Galena white	(*Berthierite*)	$FeSb_2S_4$	117
		Galena white	(*Lengenbachite*)	$Pb_6(AgCu)_2As_4S_{13}$	99
		Galena white	(*Rathite*)	$Pb_4(AsSb)S$	99
		Grayish white	(*Baumhauerite*)	$Pb_4As_6S_{13}$	117
		Grayish white	**(Molybdenite)**	MoS_2	97
		Gray	(*Erythrozincite*)	$(MnZn)S$	116

DETERMINATIVE TABLES

See summary of reactions at upper right hand corner of each page to locate minerals between thumb tabs.

CREAM *Whitneyite* Cu_9As (Very Rare)

Microchem. HNO_3, Effervesces and turns brown; rubs gray. HCl, Tarnishes and rubs to faint gray showing structure. KCN, Tarnishes brown; rubs gray with structure. $FeCl_3$, Quickly blackens; rubs to gray. $HgCl_2$, Quickly blackens with structure; rubs light brown with structure. KOH, Tarnishes faintly.

Hardness, Medium. 3.5. Very Sectile, Gray powder when scratched.

Description. C. and Str.—Silver-white. Usually massive.

B.B. Yields arsenic coat on charcoal (Chap. IV, 2, *a*) and mirror in closed tube (IV, 2, *c*). Gives azure-blue copper chloride flame with HCl (IV, 6, *b*). Fus.—2.

CREAMY WHITE *Huntilite*[1] Ag_3As? (Very Rare)

Microchem. HNO_3, Effervesces and quickly blackens; rubs to roughened surface. HCl, Liquid scarcely affects surface, but fumes tarnish persistently.

$FeCl_3$, Tarnishes iridescent; rubs to faint iridescence. KOH, Neg.

Hardness, Medium. KCN, Slowly turns dark; rubs to clean roughened surface.

Description. C.—Dark gray to black. Described from Silver Islet, Lake Superior, as massive in occurrence.

B.B. Yields arsenic coat on charcoal (IV, 2, *a*) and mirror in closed tube (IV, 2, *c*). When reduced with the fluxes yields metallic silver (IV, 18, *a*).

GRAYISH WHITE **Cuprite** Cu_2O

Microchem. HNO_3, Effervesces and forms a deposit of metallic copper; washes to coat of metallic copper; rubs gray. Fumes tarnish. HCl, Quickly darkens, forming a white coating as seen by inclined light; rubs faint. Fumes tarnish. KCN, Develops structure; rubs grayish. $FeCl_3$, Tarnishes; rubs to iridescence. Fumes tarnish. $HgCl_2$, Neg. KOH, Neg.

Hardness, Medium. 3.5–4. Slightly brittle, Blood red powder when scratched. Internal reflection seen by inclined light is deep red.

Description. C.—Red to nearly black. Str.—Shades of red or brown. With metallic copper it is found in the oxidized zone of all copper deposits.

B.B. In R.F. on charcoal yields metallic copper and colors the flame an emerald-green. Fus.—3.

[1] The microchemical reactions for minerals marked with an asterisk are from MURDOCH, JOSEPH. "The Microscopical Determination of the Opaque Minerals." John Wiley & Sons. 1916.

CREAMY WHITE *Chilenite** Ag_6Bi (Very Rare)

Microchem. HNO$_3$, Effervesces and blackens; rubs to roughened surface. HCl, Turns brown; rubs to iridescent roughened surface. KCN, Turns brownish; rubs clean. FeCl$_3$, Tarnishes iridescent; rubs gray. KOH, Neg.

Hardness, Low.

Description. C.—Silver-white, tarnishing to yellowish. Described as amorphous and granular.

B.B. When reduced with the fluxes yields metallic silver (IV, 18, *a*). With KI & S flux yields brick-red bismuth coat (IV, 3, *a*).

CREAMY WHITE *Domeykite* Cu_3As (Very Rare)

Microchem. HNO$_3$, Quickly effervesces and blackens. HCl, Most specimens develop structure. KCN, Some portions tarnish faintly, some practically negative. FeCl$_3$, Quickly tarnishes; rubs to roughened surface. HgCl$_2$, Tarnishes brown; rubs faint. KOH, Quickly tarnishes; rubs clean.

Hardness, Low to medium. 3–3.5.

Description. C.—Steel gray. Str.—Gray. Occurs reniform, botroidal, massive, and disseminated.

B.B. Yields arsenic coat on charcoal (IV, 2, *a*). Gives azure-blue copper chloride flame with HCl (IV, 6, *b*).

PURPLE *Rickardite* Cu_4Te_3 (Very Rare)

Microchem. HNO$_3$, Blackens with violent effervescence. HCl, Turns pale bluish and dissolves to pitted surface. KCN, Bleaches to a pale tint. FeCl$_3$, Bleaches to brownish. HgCl$_2$, Bleaches to bluish-green. KOH, Tarnishes iridescent.

Hardness, Low. 3.5. Brittle.

Description. C. and Str.—Purple. Massive; fracture irregular.

B.B. On charcoal fuses easily to a brittle globule, yielding a pale azure-blue flame tinged with green, and a white tellurium coat. Fus.—1.

PINK	**Native Copper**	See page 51.
BLUISH WHITE	**Chalcocite**	See page 51.
GALENA WHITE	HESSITE	See page 66.

HNO$_3$–E
HCl
KCN
Low
FeCl$_3$–N

CREAMY WHITE *Aikinite* See page 63.

REMOVE
THIS
WRITING

GRAYISH WHITE ALABANDITE MnS

Microchem. HNO_3, Effervesces and tarnishes; rubs gray, showing solution pits. Fumes tarnish brown. HCl, Nearly the same as with HNO_3. KCN, Neg. $FeCl_3$, Neg. $HgCl_2$, Fumes tarnish faintly; washes to pale brown; rubs clean. KOH, Neg.

Hardness, Medium. 3.5–4. Sectile, but slightly brittle. Olive green powder when scratched.

Internal reflection seen by inclined light is green.

Description. C.—Iron black. Str.—Green. Has perfect cubic cleavage.

B.B. Shows manganese with the sodium carbonate bead (IV, 11, *a*). Fus.—3.

WHITE *Naumannite** $(Ag_2Pb)Se$ (Very Rare)

Microchem. HNO_3, Quickly effervesces and blackens; rubs to roughened surface. HCl, Tarnishes slowly to iridescence. KCN, Neg. $FeCl_3$, Tarnishes slowly brownish; rubs clean. KOH, Neg.

Hardness, Low. 2.5.

Description. C.—Iron-black. Str.—Black. Massive with cubic cleavage.

B.B. Selenium odor and coat on charcoal (IV, 17, *a* and *b*). With KI & S flux yields lemon-yellow lead coat on charcoal (IV, 10, *a*). Yields silver button on cupellation (IV, 18, *a*). Fus.—2.

WHITE *Altaite* PbTe (Very Rare)

Microchem. HNO_3, Effervesces and quickly darkens, developing crystals and tree-like forms resembling microlites; rubs to grayish etched surface. Fumes tarnish iridescent. HCl, Quickly browns; rubs to pale brown with white patches. KCN, Neg. $FeCl_3$, Quickly tarnishes brownish iridescent; rubs to gray, roughened surface. $HgCl_2$, Neg. KOH, Neg.

Hardness, Low. 3. Sectile. Gray powder when scratched.

Description. C.—Tin-white. Str.—Gray. Has cubic cleavage.

B.B. Yields tellurium flame and coat on charcoal (IV, 20, *a*). With KI & S flux gives a lemon-yellow lead coat on charcoal. Fus.—1.5.

CREAMY WHITE **Native Silver** Ag

Microchem. HNO_3, Effervesces slightly for a moment; rubs to gray, roughened surface. Fumes tarnish. HCl, Fumes tarnish slightly; rubs clean. KCN, Neg. $FeCl_3$, Tarnishes iridescent; rubs to iridescence. $HgCl_2$, Tarnishes gray; rubs to brownish gray. KOH, Neg.

Hardness, Low. 2.5–3. Very sectile.

Description. C.—Silver-white. Str.—Silver-white. Usually in arborescent, sheet, and wire forms.

B.B. Fuses easily on charcoal and yields a brown coat of silver oxide. (For other tests see IV, 18, *a*.) Fus.—2.

CREAMY WHITE *Native Bismuth* Bi (Very Rare)

Microchem. HNO_3, Quickly effervesces and darkens; rubs to brownish gray. HCl, Slowly darkens; Fumes tarnish brown. KCN, Neg. $FeCl_3$, Darkens quickly. $HgCl_2$, Slowly deep brown; rubs to brown. KOH, Neg.

Hardness, Low. 2–2.5. Very Sectile.

Description. C.—Reddish silver-white. Str.—Silver-white. Arborescent.

B.B. On charcoal fuses easily and volatilizes completely, yielding yellow sublimate. With KI & S flux gives brick-red coat on charcoal.

CREAMY WHITE *Tapalpaite** $3Ag_2(STe).Bi_2(STe)_3$? (Very Rare)

Microchem. HNO_3, Quickly effervesces and darkens; rubs to dark, rough surface. Fumes tarnish. HCl, Faintly tarnished; rubs clean. Fumes same. KCN, Neg. $FeCl_3$, Quickly darkens; rubs to gray, roughened surface. KOH, Neg.

Hardness, Low.

Description. C.—Pale steel-gray. Str.—Gray. Massive.

B.B. Yields tellurium flame and coat on charcoal (IV, 20, *a*). With KI & S flux gives brick-red bismuth coat on charcoal (IV, 3, *a*). Fus.—1.

GALENA WHITE *Plagionite* $5PbS.4Sb_2S_3$ (Very Rare)

Microchem. HNO_3, Slowly darkens with slight effervescence; rubs to light gray. HCl, Acid negative, but fumes tarnish bright brown. KCN, Neg. $FeCl_3$, Turns brown; rubs pale brown. $HgCl_2$, Neg. KOH, Slowly iridescent; rubs clean.

Hardness, Low. 2.5. Brittle. Gray powder when scratched.

Description. C.—Blackish gray. Str.—Black.

B.B. Yields antimony coat and sublimates (IV, 1, *a–d*). With KI & S flux gives lemon-yellow lead coat (IV, 10, *a*). Fus.—1.

GALENA WHITE **Galena** See page 77.

GALENA WHITE *Semseyite* $7PbS.3Sb_2S_3$ (Very Rare)

Microchem. HNO_3, Slowly effervesces and blackens; rubs gray. Fumes tarnish brown. HCl, Slowly faint brown; rubs clean. Fumes tarnish. KCN, Neg. $FeCl_3$, Neg. $HgCl_2$, Neg. KOH, Some specimens unaffected; others faintly tarnish brown.

Hardness, Low. Brittle. Gray powder when scratched.

Description. C.—Gray. Str.—Black. Tabular.

B.B. Yields antimony coat and sublimates (IV, 1, *a–d*). With KI & S flux gives lemon-yellow coat (IV, 10, *a*). Fus.—1.

WHITE	*Aikinite*	See page 63.
GALENA WHITE	*Plagionite*	See page 43.
GALENA WHITE	*Dognacskaite*	See page 63.
GRAYISH WHITE	JAMESONITE	See page 62.
GRAYISH WHITE	ALABANDITE	See page 41.
GALENA WHITE	*Horsfordite*	See page 63.

VIOLET WHITE *Polydymite* See page 57.

REMOVE
THIS
WRITING

HNO_3–E
HCl–N
KCN
Med.
$FeCl_3$

PINK	**Native Copper**	See page 51.

REMOVE
THIS
WRITING

PINK **Native Copper** Cu

Microchem. HNO_3, Effervesces without tarnishing; rubs to roughened surface. HCl, Neg. (Sometimes tarnishes faintly.) KCN, Slowly browns; rubs to clean roughened surface. $FeCl_3$, Quickly darkens and dissolves. $HgCl_2$, Quickly blackens; rubs to iridescence. KOH, Slowly turns brown to bluish; rubs bluish.

Hardness, Medium. 2.5–3. Sectile. Pink when scratched.

Description. C.—Copper-red. Str.—Shining. Malleable, ductile. Shows arborescent and irregular structures. Occurs in oxidized zone of all copper deposits.

B.B. Easily fusible, yielding an emerald-green flame in the O.F. On charcoal becomes black after fusion.

BLUISH WHITE **Chalcocite** Cu_2S

Microchem. HNO_3, Effervesces vigorously, turning blue, and developing cracks or cleavage; rubs same. HCl, Tarnishes very slightly. KCN, Quickly blackens; rubs to etched surface, with structure. $FeCl_3$, Tarnishes slightly. $HgCl_2$, Tarnishes slightly. KOH, Neg.

Hardness, Low. 2.5–3. Very Sectile. Gray powder when scratched.

Description. C. and Str.—Dark lead-gray, tarnishing to blue on exposure. Occurs in zone of secondary enrichment in all copper deposits.

B.B. Fuses easily and boils with spirting. Powdered and roasted without fusing, then heated in R.F., yields metallic copper. Fus.—2–2.5.

PINKISH BROWN **Bornite** **See page 53.**

PINKISH BROWN **Bornite** Cu_5FeS_4

Microchem. HNO_3, Effervesces and turns yellowish brown; rubs to brown etched surface. Fumes tarnish. HCl, Neg. KCN, Browns; rubs to dark etched surface. Fumes tarnish. $FeCl_3$, Neg. $HgCl_2$, Neg. KOH, Neg.

Hardness, Low. 3. Sectile. Golden-brown powder when scratched.

Description. C.—Copper-red to brown; purplish bronze from tarnish. Str.—Grayish black. Massive. Common in most copper deposits.

B.B. Fuses easily on coal in the R.F. to a magnetic globule. Roasted and reduced with sodium carbonate, yields malleable copper buttons (IV, 6, *a*). Fus.—2.

CREAMY WHITE *Aikinite* See page 63.

GALENA WHITE **Stibnite** See page 87.

CREAMY PINK **Niccolite** NiAs

Microchem. HNO_3, Effervesces and tarnishes with etching; rubs to gray, etched surface. Fumes tarnish. HCl, Neg. KCN, Neg. $FeCl_3$, Slowly tarnishes brown; rubs to pale brown. (Some specimens show this reaction very weakly.) $HgCl_2$, Tarnishes brown; washes to persistent brown. KOH, Neg.

Hardness, High. 5–5.5. Very brittle. Gray powder when scratched.

Description. C.—Pale copper-red. Str.—Brownish. Occurs massive and disseminated.

B.B. Fuses easily on coal to brittle globule, yielding arsenical odor and possibly a white coat of oxide. With dimethyl glyoxime gives a red precipitate (IV, 14, *b*).

PINKISH WHITE *Maucherite* Ni_3As_2 (Very Rare)

Microchem. HNO_3, Effervesces and blackens quickly; rubs to gray, rough surface. HCl, Neg. KCN, Neg. $FeCl_3$, Slowly tarnishes brownish; rubs faint. $HgCl_2$, Faintly browns; rubs clean. KOH, Neg.

Hardness, High. 5. Very brittle. Gray powder when scratched.

Description. C.—Reddish silver-white, tarnishing gray copper-red. Str.—Blackish gray.

B.B. Same as niccolite.

(NOTE: Temiskamite is not a mixture, but a homogeneous mineral identical with Maucherite.)

WHITE **Smaltite** $CoAs_2$

Microchem. HNO_3, Quickly effervesces and darkens; rubs to gray, showing lath structure. Fumes tarnish. HCl, Neg. KCN, Neg. $FeCl_3$, Slowly tarnishes brownish, showing structure; rubs pale. (This reaction is not always decisive.) $HgCl_2$, Slowly browns; rubs clean. KOH, Neg.

Hardness, High. 5.5–6. Brittle.

Description. C.—Tin-white. Str.—Gray-black. Occurs massive.

B.B. Easily fusible on charcoal, yielding a magnetic globule and an arsenical odor; may yield a white arsenic coat. With borax shows a persistent cobalt blue (IV, 5, *a*). Fus.—2.5.

(*Safflorite* is the same as smaltite.)

WHITE **Arsenopyrite** FeAsS

Microchem. HNO_3, Slowly effervesces and quickly tarnishes through iridescence to deep brown; rubs to roughened surface. HCl, Neg. KCN, Neg. $FeCl_3$, Neg. $HgCl_2$, Neg. KOH, Neg.

Hardness, High. 5.5–6. Brittle. Gray powder when scratched.

Description. C.—Silver-white to steel-gray. Str.—Grayish black. Occurs as orthorhombic crystals, massive, granular, or compact.

B.B. On charcoal fuses easily to brittle magnetic globule and evolves arsenic odor. In the closed tube it yields first an arsenious sulphide sublimate and then an arsenic mirror. Fus.—2.

VIOLET WHITE *Polydymite* Ni_4S_5 (Very Rare)

Microchem. HNO_3, Slowly effervesces and turns to brownish or bluish; rubs to brown or blue, showing structure. HCl, Neg. (Acid turns green, but surface washes clean.) KCN, Neg. $FeCl_3$, Neg. $HgCl_2$, Neg. KOH, Neg.

Hardness, High. 4.5. Brittle.

Description. C.—Steel-gray. Str.—Grayish black.

B.B. With dimethyl glyoxime yields red precipitate (IV, 14, *b*). Fus.—2.

WHITE *Willyamite** (CoNi)SbS (Very Rare)

Microchem. HNO_3, Slowly effervesces and turns dark brown; rubs gray, showing etching. HCl, Neg. KCN, Neg. $FeCl_3$, Neg. KOH, Neg.

Hardness, High. 5.5.

Description. C.—Tin-white to steel-gray. Str.—Grayish black. Perfect cubic cleavage. Massive.

B.B. Tests for cobalt, nickel, and antimony in Chapter IV.

WHITE *Kallilite** Ni(SbBi)S? (Very Rare)

Microchem. Like Willyamite.

Hardness, High to medium.

Description. C.—Light bluish gray. Str.—Black. Massive.

B.B. Tests for nickel, antimony, and bismuth in Chapter IV.

CREAMY WHITE **Pyrite** FeS_2

Microchem. HNO_3, Very slowly effervesces and faintly browns; rubs to faint brown. Fumes tarnish slowly. Other reagents negative.

Hardness, High. 6–6.5. Cannot be scratched.

Description. C.—Light brass-yellow. Str.—Greenish to brownish black. Pyrite is the most common of all sulphides and usually the oldest.

B.B. Becomes magnetic on charcoal in the R.F. Fus.—3.

CREAMY WHITE **Marcasite** FeS_2

Microchem. Like Pyrite.

Hardness, High. 6–6.5.

Description. Like Pyrite except in crystal form and in occurrence. It occurs always as a secondary mineral.

B.B. Like Pyrite.

CREAMY PINK **Niccolite** See page 55.

CREAMY WHITE *Cosalite* $Pb_2Bi_2S_5$ (Very Rare)

Microchem. HNO_3, Instantly effervesces and blackens; rubs to dark gray. Fumes tarnish. HCl, Neg. (Sometimes slowly faint tarnish; rubs clean.) KCN, Neg. $FeCl_3$, Very slowly faint brown; sometimes appears negative. $HgCl_2$, Fumes tarnish; rubs clean. KOH, Tarnishes brown to iridescent.

Hardness, Medium to low. 2.5–3. Brittle. Gray powder when scratched.

Description. C.—Lead-gray. Str.—Grayish black.

B.B. With KI & S flux shows both lead and bismuth (IV, 3 and 10). Fus.—1.5.

GALENA WHITE *Native Arsenic* As (Very Rare)

Microchem. HNO_3, Instantly blackens with slow effervescence; rubs to brownish gray. HCl, Neg. KCN, Neg. $FeCl_3$, Quickly darkens to persistent brownish. $HgCl_2$, Slowly tarnishes pale brown; rubs faint. KOH, Neg.

Hardness, Medium to low. 3.5. Sectile, but slightly brittle. Gray powder when scratched.

Description. C.—Tin-white. Str.—Gray. Has a basal cleavage.

B.B. Yields white arsenic oxide coat on coal and easily volatilizes without fusion.

CREAM *Melonite** Ni_2Te_3 (Very Rare)

Microchem. HNO_3, Quickly effervesces and blackens; rubs gray. HCl, Neg. KCN, Neg. $FeCl_3$, Slowly tarnishes; rubs faint. KOH, Neg.

Hardness, Low.

Description. C.—Reddish white. Str.—Gray. Has basal cleavage.

B.B. Easily fusible, but not entirely volatile, yielding white coat on charcoal and coloring flame bright green (IV, 20, *a*). With dimethyl glyoxime gives a red precipitate (IV, 14, *b*).

CREAMY WHITE CALAVERITE $AuTe_2 + Ag$

Microchem. HNO_3, Slowly effervesces and turns brown; rubs to lighter brown with etching. HCl, Neg. KCN, Neg. $FeCl_3$, Slowly tarnishes brown; rubs to faint brown. $HgCl_2$, Neg. KOH, Tarnishes pale brown; rubs to faint brown.

Hardness, Low. 2.5. Brittle. Gray powder when scratched.

Description. C.—Silver-white. Str.—Gray.

B.B. Gives white coat on charcoal and colors flame bright green. Roasted and reduced with sodium carbonate, yields gold and silver; sometimes reduces to gold bead easily without soda. Fus.—1.

WHITE *Native Tellurium* Te (Very Rare)

Microchem. HNO_3, Quickly effervesces and blackens; rubs to gray, rough surface. HCl, Neg. KCN, Neg. $FeCl_3$, Slowly tarnishes iridescent. KOH, Neg.

Hardness, Low. 2–2.5. Somewhat sectile. Gray powder when scratched.

Description. C.—Tin-white. Str.—Gray. Has prismatic cleavage.

B.B. Easily fusible and volatile, yielding a white coat on charcoal and coloring the flame bright green (IV, 20, *a*). Fus.—1.

GALENA WHITE *Rezbanyite* $4PbS.5Bi_2S_3$ (Very Rare)

Microchem. HNO_3, Effervesces and tarnishes iridescent; rubs to gray etched surface. Fumes tarnish. HCl, Neg. KCN, Neg. $FeCl_3$, Slowly faint brown. $HgCl_2$, Neg. KOH, Neg.

Hardness, Low. 2.5–3. Brittle. Gray powder when scratched.

Description. C.—Lead-gray. Str.—Grayish black. Massive.

B.B. With KI & S flux shows both lead and bismuth (IV, 3 and 10). May be isomorphous with small amounts of copper and silver.

GALENA WHITE *Tetradymite* $Bi_2(Te,S)_3$

Microchem. HNO_3, Quickly tarnishes with vigorous effervescence; rubs iridescent. HCl, Neg. KCN, Neg. $FeCl_3$, Tarnishes slightly, developing scratches. $HgCl_2$, Neg. KOH, Neg.

Hardness, Low. 1.5–2. Slightly sectile.

Description. C.—Steel-gray, splendent. Str.—Gray. Perfect basal cleavage. Soils paper.

B.B. On charcoal fuses, gives white fumes and entirely volatilizes; tinges the R. F. bluish-green; coats the charcoal at first white and finally orange yellow

CREAMY WHITE **KRENNERITE** See page 63.

CREAMY WHITE **Native Silver** See page 43.

GALENA WHITE ***Plagionite*** See page 43.

GRAYISH WHITE ***Petzite*** See page 97.

Galena White *Bismuthinite* Bi_2S_3 (Very Rare)

Microchem. HNO_3, Blackens with slow effervescence; rubs to gray, roughened surface. HCl, Neg. KCN, Neg. $FeCl_3$, Neg. $HgCl_2$. Tarnishes brown; rubs to faint brown. KOH, Neg.

Hardness, Low. 2. Slightly brittle. Gray powder when scratched.

Description. C.—Lead-gray.

B.B. With KI & S flux gives a brick-red bismuth coat on charcoal (IV, 3, *a*). Fus.—1.

Grayish White Realgar AsS

Microchem. HNO_3, Effervesces without visible change. HCl, Neg. KCN, Neg. $FeCl_3$, Neg. $HgCl_2$, Neg. KOH, Tarnishes brown to black with solution.

Hardness, Low. 1.5–2.

Internal reflection seen by inclined light is orange.

Description. C.—Dark orange-red. Str.—Lighter. Occurs with antimony, arsenic, and silver ores.

B.B. On charcoal in the O.F., burns, yielding arsenic odor and no residue when pure. In closed tube yields cherry-red sublimate of arsenic sulphide. Fus.—1.

Grayish White Jamesonite $2PbS.Sb_2S_3$

Microchem. HNO_3, Quickly brown to black with very slow effervescence (Eff. often not detected); rubs to iridescent gray. HCl, Neg. Fumes tarnish slowly; rubs faint. KCN, Neg. $FeCl_3$, Neg. $HgCl_2$, Neg. KOH, Slowly develops grain structure; some grains are iridescent, others gray.

Hardness, Low. 2–3. Brittle. Gray powder when scratched.

Description. C.—Steel-gray. Str.—Grayish black. Perfect basal cleavage. Occurs fibrous or massive.

B.B. Yields antimony and lead coat on charcoal (IV, 1 and 10). With KI & S flux gives lemon-yellow lead coat on charcoal (IV, 10, *a*). Fus.—1.

Grayish White *Zinkenite* $PbS.Sb_2S_3$ (Very Rare)

Microchem. Like Jamesonite.

Hardness, Low. 3–3.5.

Description. C. and Str.—Steel-gray. Occurs in columnar crystals, striated lengthwise, also fibrous and massive.

B.B. Like Jamesonite.

Bluish Gray *Tungstenite* WS_2(?) (Very Rare)

Microchem. HNO_3, Does not change color, but effervesces after a few moments. HCl, Neg. KCN, Neg. $FeCl_3$, Neg. $HgCl_2$, Quickly tarnishes brown; rubs clean. KOH, Neg.

Hardness, Low. 2.5.

Description. C.—Dark lead-gray. Str.—Dark gray. Marks paper.

B.B. See tests for tungsten (IV, 25).

White *Native Tellurium* **See page 61.**

Galena White **Stibnite** **See page 87.**

CREAMY WHITE KRENNERITE $AuTe_2 + Ag$

Microchem. HNO_3, Slowly effervesces and turns brown; rubs to lighter brown with etching. HCl, Neg. KCN, Neg. $FeCl_3$, Neg. $HgCl_2$, Neg. KOH, Tarnishes pale brown; rubs to faint brown.

Hardness, Low to medium. 2.5. Sectile, but slightly brittle. Gray powder when scratched.

Description. C.—Silver-white. Str.—Gray. Has perfect basal cleavage. (Calaverite lacks this perfect cleavage.)

B.B. Gives white coat on coal and colors flame bright green. Roasted and reduced with sodium carbonate yields gold and silver; sometimes reduces to gold bead easily without soda. Fus.—1.

CREAMY WHITE *Emplectite* $CuBiS_2$ (Very Rare)

Microchem. HNO_3, Effervesces slightly and tarnishes pale brown; rubs clean. Fumes tarnish faintly. HCl, Neg. KCN, Neg. $FeCl_3$, Neg. $HgCl_2$, Neg. KOH, Tarnishes very slowly brown; rubs clean.

Hardness, Low. 2. Slightly brittle. Gray powder when scratched.

Description. C.—Grayish white. Str.—Black.

B.B. Yields azure-blue copper chloride flame with HCl (IV, 6, *b*). With KI & S flux gives brick-red bismuth coat on charcoal (IV, 3, *a*). Fus.—1.

CREAMY WHITE *Chiviatite** $Pb_2Bi_2S_{11}$ (Very Rare)

Microchem. HNO_3, Slowly effervesces and turns iridescent to gray; rubs to clean, somewhat roughened surface. HCl, Neg. KCN, Neg. $FeCl_3$, Neg. KOH, Neg.

Hardness, Low.

Description. C.—Lead-gray. Str.—Grayish black. Commonly foliated.

B.B. With KI & S flux shows both lead and bismuth (IV, 3 and 10).

CREAMY WHITE *Aikinite* $3(PbCu_2)S.Bi_2S_3$ (Very Rare)

Microchem. HNO_3, Effervesces and blackens; rubs to heavy gray. Fumes tarnish brown. HCl, Neg. KCN, Neg. (Some specimens very slowly browned; rubs clean.) $FeCl_3$, Neg. $HgCl_2$, Neg. KOH, Neg.

Hardness, Low. 2–2.5. Brittle. Gray powder when scratched.

Description. C.—Lead-gray. Str.—Grayish black.

B.B. Gives azure-blue copper chloride flame with HCl (IV, 6, *b*). With KI & S flux shows both lead and bismuth (IV, 3 and 10). Fus.—1–1.5.

GALENA WHITE *Dognacskaite* Cu, Bi, and S (Very Rare)

Microchem. HNO_3, Quickly effervesces and turns iridescent to black; rubs to gray, roughened surface. Fumes tarnish. HCl, Neg. Fumes tarnish brown; rubs to pale brown. KCN, Neg. $FeCl_3$, Neg. $HgCl_2$, Neg. KOH, Neg.

Hardness, Low. Very Sectile. Gray powder when scratched.

Description. C.—Gray, tarnishing on exposure to the air.

B.B. Same as Emplectite.

GALENA WHITE *Boulangerite* $3PbS.Sb_2S_3$ (Very Rare)

Microchem. HNO_3, Effervesces and blackens; washes to etched structure. Fumes tarnish light brown. Other reagents negative.

Hardness, Low. 2.5–3. Sectile, but slightly brittle. Gray powder when scratched.

Description. C.—Bluish lead-gray. Str.—Black. Occurs fibrous, granular, or compact.

B.B. Like Jamesonite.

GALENA WHITE *Horsfordite* Cu_6Sb (Very Rare)

Microchem. HNO_3, Slightly effervesces and quickly blackens; rubs to faint gray. HCl, Practically negative; some specimens turn very slowly faint brown and rub to very faint gray. KCN, Neg. $FeCl_3$, Neg. $HgCl_2$, Tarnishes a brown rim which rubs clean. KOH, Slowly tarnished faintly.

Hardness, Low. Very Sectile. Gray powder when scratched.

Description. C.—Silver-white. Occurs massive.

B.B. Reacts for antimony and copper. Fus.—1.5.

CREAMY WHITE *Sulvanite* Cu_3VS_4 (Very Rare)
Microchem. HNO_3, Neg. Fumes tarnish; rubs clean. HCl, Neg. Fumes tarnish; rubs clean. KCN, Neg. Fumes tarnish; rubs clean. $FeCl_3$, Neg. $HgCl_2$, Neg. KOH, Neg.
Hardness, Medium. 3.5. Brittle. Gray powder when scratched
Description. C.—Bronze-yellow. Occurs massive. (South Australia.)
B.B. Roasted and reduced with sodium carbonate and borax yields copper buttons. Gives vanadium bead with the fluxes (IV, 27, *a*).

GALENA WHITE *Guejarite* See page 83.

Galena White *Hessite* Ag_2Te

Microchem. HNO_3, Slowly tarnishes iridescent to brown; rubs to roughened surface. (Sometimes slowly effervesces.) HCl, Tarnishes iridescent; rubs faint or clean. (Sometimes nearly negative.) KCN, Slowly tarnishes. $FeCl_3$, Tarnishes iridescent; rubs faint to clean. $HgCl_2$, Tarnishes brown; rubs clean. KOH, Very slowly tarnishes slightly; rubs faint.

Hardness, Low. 2.5–3. Quite sectile.

Description. C.—Steel-gray. Str.—Gray. Occurs crystalline, massive, etc.

B.B. Coat and flame are not decisive. Shows tellurium with concentrated H_2SO_4. Reduced with sodium carbonate, yields silver buttons. Fus.—1.

Brown **Pyrolusite** **See page 101.**

CREAMY WHITE *Dyscrasite* Ag_3Sb, Ag_6Sb, etc. (Very Rare)

Microchem. HNO_3, Tarnishes and develops differential etching. HCl, Slowly tarnishes with development of differential etching. KCN, Slowly and faintly etches differentially. $FeCl_3$, Differential iridescent tarnish. $HgCl_2$, Tarnishes to brownish iridescence with structure; rubs same. Fumes tarnish light brown. KOH, Neg.

Hardness, Low. 3.5–4. Exceedingly sectile.

Description. C.—Silver-white, sometimes tarnishing yellow or blackish. Str.—Silver-gray.

B.B. Easily fuses to a globule, coating the charcoal with white antimony trioxide, and finally leaving a globule of pure silver. Soluble in nitric acid. Fus.—1.5.

GRAYISH WHITE **Argentite** Ag_2S

Microchem. HNO_3, Slowly tarnishes light brown; rubs clean. Fumes tarnish brown. HCl, Sometimes slowly tarnishes faint brown, but often almost negative. KCN, Tarnishes brown; rubs to faint gray. Fumes tarnish brown. $FeCl_3$, Slowly brown to iridescent; rubs to very faint iridescence. $HgCl_2$, Tarnishes brown; rubs to pale brown. KOH, Neg. Blackened by short exposure to unscreened strong electric arc light.

Hardness, Low. 2–2.5. Exceedingly sectile. Black powder when scratched.

Description. C.—Lead-gray. Str.—Shining gray. Common in many silver deposits and also occurs in disseminated small crystals in silver bearing lead minerals, etc.

B.B. Fuses with intumescence in the O.F. on charcoal, emitting sulphur dioxide odor and a globule of pure silver. Fus.—1.5.

GRAYISH WHITE *Stromeyerite* $(Ag,Cu)_2S$ (Very Rare)

Microchem. HNO_3, Tarnishes slightly with etching; rubs to rough gray. Fumes tarnish. HCl, Slowly very faint brown. (Sometimes practically negative.) Fumes tarnish. KCN, Tarnishes brown; rubs to faint gray. $FeCl_3$, Quickly tarnishes brown or iridescent with structure; rubs to brownish iridescence. $HgCl_2$, Tarnishes brown; rubs to pale yellowish. KOH, Neg.

Hardness, Low. 2.5–3. Very sectile. Black powder when scratched.

Description. C. and Str.—Dark steel-gray. Prismatic, massive, compact.

B.B. Fuses in the O. F. on charcoal to a semi-malleable globule, which reacts for copper (IV, 6) and cupelled with lead yields a silver button. Soluble in nitric acid. Fus.—1.5.

GRAYISH WHITE *Jalpaite* $3Ag_2S.Cu_2S$? (Very Rare)

Microchem. HNO_3, Acid has no effect, but the fumes tarnish slowly brown; rubs clean. HCl, Acid turns green and surface is slightly roughened, but the reaction often appears negative. KCN, Tarnishes brown very slowly; rubs clean. $FeCl_3$, Tarnishes and rubs to faint greenish iridescence. $HgCl_2$, Tarnishes brown; rubs to faint spot. Fumes tarnish. KOH, Neg.

Hardness, Low. Specimen from Tres Puntas, Chili is brittle and has a blood red powder when scratched. Internal reflection seen with inclined light is blood red.

Description. C.—Blackish lead-gray.

B.B. Like stromeyerite.

BLUISH WHITE *Brongniardite** $PbS.Ag_2S.Sb_2S_3$? (Very Rare)

Microchem. HNO_3, Slowly tarnishes pale brown; rubs clean. HCl, Slowly tarnishes faint brown; rubs clean. KCN, Slowly tarnishes faint brown; rubs clean. $FeCl_3$, Neg. KOH, Tarnishes iridescent; rubs to clean etched surface.

Hardness, Low. 3+.

Description. C. and Str.—Grayish black. Massive.

B.B. On charcoal, decrepitates, fuses easily, and gives off sulphur dioxide vapors. Upon roasting yields an impure globule of silver and a yellow coating of lead oxide. In the closed tube yields a faint orange sublimate. In the open tube, fuses, and yields a sublimate of white antimony trioxide. Fus.—1.

GRAY PSILOMELANE H_4MnO_5?

Microchem. HNO_3, Acid tarnishes faintly; fumes strongly. Rubs nearly clean. HCl, Tarnishes and rubs faint brown. KCN, Neg. $FeCl_3$, Neg. $HgCl_2$, Neg. KOH, Neg.

Hardness, High. 5–6. Brittle.

Description. C.—Iron-black. Str.—Shining brownish black. Massive, botroidal, reniform, and stalactitic.

B.B. Infusible. In the closed tube yields water. Yields a green bead with sodium carbonate (IV, 11, *a*). Soluble in HCl with evolution of chlorine. May contain enough iron to become magnetic when roasted.

GRAYISH WHITE *Tenorite* CuO (Very Rare)

Microchem. HNO_3, Acid without effect, but fumes tarnish; washes clean. HCl, Tarnishes and deposits a mass of white acicular crystals; rubs to faint gray. Fumes tarnish. KCN, Neg. $FeCl_3$, Tarnishes pale brown very slowly; rubs clean. $HgCl_2$, Neg. KOH, Neg.

Hardness, Medium. 3–4. Slightly brittle.

Description. C.—Iron-gray to black. Str.—Black. Occurs massive, crusted, in minute crystals, and scales. Found in oxidation zone in some Cu deposits.

B.B. Like cuprite (page 33).

HNO_3
HCl
KCN-N
Med.
$FeCl_3$ -N

GRAY *Cuprodescloizite* (Pb, Zn, Cu)$_4$$V_2O_9$.$H_2O$? (Very Rare)

Microchem. HNO_3, Blackens; rubs to roughened surface. Fumes tarnish brown. HCl, Tarnishes with structure, acid turning yellow; rubs to slightly roughened surface. Fumes tarnish. KCN, Neg. $FeCl_3$, Neg. $HgCl_2$, Neg. KOH, Neg.

Hardness, Medium. 3.5. Brittle. Reddish yellow powder when scratched. Internal reflection as seen by inclined light is greenish or yellowish brown, but may not always be noticeable.

Description. C.—Dull green to greenish black or yellowish brown. Usually in crusts or reniform masses with mammillary surface and columnar structure.

B.B. In the closed tube yields water. Will test for constituents in the formula above as well as for arsenic and other impurities. (See Chap. IV.)

GRAYISH WHITE TENORITE See page 73.

GRAY **Sphalerite** **See page 94.**

GRAYISH WHITE *Aguilarite* $Ag_2S.Ag_2Se$ (Very Rare)

Microchem. HNO_3, Slowly tarnishes brown; rubs clean. Fumes tarnish. HCl, Acid without effect, but fumes often tarnish brown; rubs to faint brown. KCN, Neg. $FeCl_3$, Slowly tarnishes iridescent; rubs to faint iridescence. $HgCl_2$, Tarnishes brown to iridescent; rubs to iridescence. KOH, Neg.

Hardness, Low. 2.5. Extremely sectile.

Description. C.—Iron-black. Str.—Black.

B.B. On charcoal yields a white coat with metallic-like luster, also a selenium odor and blue flame. Heated slowly in the open tube yields metallic silver. Fus.—1.

PURPLE *Umangite** Cu_3Se_2 (Very Rare)

Microchem. HNO_3, Turns blue; rubs same. HCl, Same as HNO_3.- KCN, Neg. $FeCl_3$, Same as HNO_3. KOH, Slowly tarnishes brown; rubs blue.

Hardness, Low. 3.

Description. C.—Cherry-red, tarnishing to violet-blue. Str.—Black. Massive.

B.B. Roasted and reduced with sodium carbonate on charcoal, yields copper buttons. Also yields a white coat with metallic-like luster, a selenium odor, and colors the flame blue (IV, 17, *a*). Fus.—1.5.

WHITE *Clausthalite** PbSe (Very Rare)

Microchem. HNO_3, Forms brick-red coating; rubs off to brown surface. HCl, Slowly tarnishes brown; rubs clean. KCN, Neg. $FeCl_3$, Slowly tarnishes, forming bluish and yellowish coating. KOH, Neg.

Hardness, Low. 2.5–3.

Description. C.—Lead-gray. Str.—Black. Granular masses; cubic cleavage.

B.B. Decrepitates in the closed tube. Gives all tests for lead and selenium. (See Chap. IV, 10 and 17.) Fus.—2.

GALENA WHITE **Galena** PbS

Microchem. HNO_3, Quickly blackens; rubs black, with rough surface. (In some specimens shows effervescence.) HCl, Tarnishes brown; rubs to gray. Fumes tarnish. KCN, Neg. $FeCl_3$, Tarnishes brown to iridescent; rubs brown. $HgCl_2$, Neg. KOH, Neg.

Hardness, Low. 2.5–2.75. Sectile, but slightly brittle. Gray powder.

Description. C. and Str.—Pure lead-gray. Perfect cubic cleavage.

B.B. Yields all tests for lead (IV, 10). Fus.—2.

GALENA WHITE *Lillianite** $3PbS.Bi_2S_3$ (Very Rare)

Microchem. HNO_3, Quickly tarnishes iridescent; rubs clean. HCl, Slowly tarnishes pale brown; rubs clean. Fumes tarnish. KCN, Neg. $FeCl_3$, Slowly tarnishes iridescent; rubs clean. KOH, Neg.

Hardness, Low.

Description. C.—Steel-gray. Str.—Blck. Massive, also crystalline.

B.B. With KI & S flux shows both lead and bismuth (IV, 3 and 10). Fus.—1–1.5.

GRAYISH WHITE *Teallite** $PbSnS_2$ (Very Rare)

Microchem. HNO_3, Tarnishes brown to iridescent; rubs clean. HCl, Slowly tarnishes light brown; rubs clean. Fumes tarnish. KCN, Neg. $FeCl_3$, Slowly tarnishes faint brown. KOH, Same as $FeCl_3$.

Hardness, Low. 1–2.

Description. C.—Blackish gray. Str.—Black. Thin flexible folia. Perfect basal cleavage.

B.B. Yields all tests for lead and tin. (See Chap. IV.)

GRAYISH WHITE *Cylindrite* $Pb_6Sb_2Sn_6S_{21}$ (Very Rare)

Microchem. HNO_3, Slowly tarnishes brown; rubs to iridescence. Fumes tarnish iridescent. HCl, Acid is without effect, but fumes tarnish brown; rubs to very faint gray. KCN, Neg. $FeCl_3$, Tarnishes brown; rubs to very faint brown. $HgCl_2$, Neg. KOH, Tarnishes brown to iridescent; rubs to faint gray.

Hardness, Low. 2.5–3. Very sectile. Gray powder when scratched.

Description. C.—Blackish gray. Str.—Black. Massive; also in cylindrical forms separating into distinct shells under pressure.

B.B. See Chapter IV for tests for antimony, tin, and lead. Fus.—1.5.

CREAMY WHITE **Native Silver** See page 43.

WHITE *Native Antimony* See page 96.

GALENA WHITE *Meneghinite** $4PbS.Sb_2S_3$ (Very Rare)

Microchem. HNO_3, Quickly tarnishes and blackens; rubs to gray, rough surface. HCl, Very slowly faint brown to almost negative. KCN, Neg. $FeCl_3$, Neg. $HgCl_2$, Neg. KOH, Tarnishes very slowly.

Hardness, Low. 2.5. Brittle. Gray powder when scratched.

Description. C.—Blackish gray. Str.—Black. Pinacoidal cleavage. Slender prismatic crystals; also massive, fibrous, and compact.

B.B. Decrepitates and fuses very easily. Yields antimony flame and coat on charcoal (IV, *a*). With KI & S flux yields lemon-yellow lead iodide coat (IV, 10 *a*). Fus.—1.

GALENA WHITE *Geocronite* $5PbS.Sb_2S_3$ (Very Rare)

Microchem. Same as Meneghinite. Except KOH is negative.

Hardness, Low. 2.5. Brittle. Gray powder when scratched.

Description. C. and Str.—Light lead-gray to grayish blue. Usually massive, granular, or earthy.

B.B. Like Meneghinite.

GRAYISH WHITE *Franckeite* $Pb_5Sn_2Sb_2S_{12}$ (Very Rare)

Microchem. HNO_3, Tarnishes brown, then iridescent; rubs faint iridescent. HCl, Usually tarnishes faint brown. Fumes tarnish slightly. KCN, Neg. $FeCl_3$, Neg. $HgCl_2$, Neg. KOH, Practically negative, but sometimes faintly tarnishes and rubs clean.

Hardness, Low. 2.75. Extremely sectile. Gray powder when scratched.

Description. C.—Blackish gray. Str.—Black. One perfect cleavage. Massive.

B.B. See Chapter IV for tests for lead, tin, antimony, etc. Fus.—1.

WHITE *Eucairite* See page 97.

GRAYISH WHITE *Cylindrite* See page 77.

HNO_3
HCl–N
KCN
Med.
$FeCl_3$

GRAYISH WHITE POLYBASITE See page 87.

GRAYISH (BROWNISH) WHITE *Freibergite* $(Cu, Ag)_8Sb_2S_7$

Microchem. HNO_3, Slowly tarnishes brown to iridescent; rubs pale. Fumes tarnish brown. HCl, Neg. KCN, Tarnishes brown to iridescent; rubs same. $FeCl_3$, Neg. $HgCl_2$, Neg. KOH, Neg.

Hardness, Medium. 3–4.

Description. Similar to tetrahedrite.

B.B. Like tetrahedrite, but contains silver.

PINKISH WHITE *Luzonite* Cu_3AsS_4 (Very Rare)

Microchem. HNO_3, Very slowly tarnishes faintly; washes clean. Fumes tarnish slightly. HCl, Neg. KCN, Slowly tarnishes. $FeCl_3$, Neg. $HgCl_2$, Neg. KOH, Neg.

Hardness, Medium. 3.5. Brittle. Gray or black powder when scratched.

Description. C.—Dark reddish steel-gray. Str.—Black. Massive.

B.B. Like enargite.

BROWNISH WHITE **Enargite** Cu_3AsS_4

Microchem. HNO_3, Very slowly tarnishes faint brown; rubs clean. Fumes slowly tarnish faint brown. HCl, Neg. KCN, Sometimes reacts very faintly, but quickly darkens some specimens, rubbing to a rough etched surface. $FeCl_3$, Neg. $HgCl_2$, Slowly tarnishes brown; rubs clean. KOH, Neg.

Hardness, Medium. 3. Brittle. Gray powder when scratched.

Description. C. and Str.—Grayish to iron-black. Often seen with prismatic planes vertically striated; also massive or disseminated. Columnar cleavage.

B.B. In the closed tube decrepitates, yielding a sublimate of sulphur. At a higher temperature it fuses and yields a sublimate of arsenic sulphide (IV, 2, *c*). Roasted and reduced with sodium carbonate, yields a malleable copper button.

PINKISH WHITE *Famatinite* Cu_3SbS_4

Microchem. HNO_3, Very slowly tarnishes brown; rubs to brown. Fumes tarnish faintly. HCl, Neg. KCN, Slowly tarnishes brown with etching; rubs to pale etched surface. (Some specimens are practically negative.) $FeCl_3$, Neg. $HgCl_2$, Neg. KOH, Neg.

Hardness, Medium. 3.5. Brittle. Gray powder when scratched.

Description. C.—Gray with tinge of copper-red. Str.—Black.

B.B. In the closed tube decrepitates, yielding sublimate of sulphur, and on higher heating, a sublimate of antimony sulphide (IV, 1, *c*). Reduced with fluxes, yields metallic copper and, when moistened with HCl, yields a blue copper chloride flame (IV, 6). Fus.—1–1.5.

GALENA WHITE *Guejarite* $Cu_2Sb_4S_7$ (Very Rare)

Microchem. HNO_3, Tarnishes dark; rubs to gray. HCl, Neg. (Some specimens tarnish very faintly and rub clean.) KCN, Slowly tarnishes brown; rubs to a faint brown. $FeCl_3$, Neg. $HgCl_2$, Tarnishes faint brown; rubs clean. KOH, Quickly tarnishes; rubs to deeply etched surface.

Hardness, Medium. 3.5. Sectile, but slightly brittle. Gray powder when scratched.

Description. C.—Steel-gray with tinge of blue. Str.—Black. Prismatic crystals. Good pinacoidal cleavage.

B.B. Like Famatinite.

YELLOW	**Chalcopyrite**	See page 117.
GRAYISH WHITE	**Tennantite**	See page 117.
GRAYISH WHITE	**Tetrahedrite**	See page 117.
CREAMY WHITE	*Sulvanite*	See page 65.

GRAYISH (GREENISH) WHITE *Pearcite* Ag_9AsS_6 (Very Rare)
Microchem. HNO_3, Acid without effect, but fumes tarnish and wash clean. HCl, Neg. KCN, Quickly blackens; rubs to gray, roughened surface. $FeCl_3$, Tarnishes iridescent; rubs to very faint gray. (Sometimes rubs clean.) $HgCl_2$, Tarnishes brown; rubs to iridescent brown. KOH, Neg.
Hardness, Low. 3. Sectile.
Description. C. and Str.—Black. Tabular crystals and massive.
B.B. Yields arsenic odor and white coat on charcoal (IV, 2, *a*). Roasted and reduced with sodium carbonate yields silver buttons. Fus.—1.

BLUISH WHITE *Polyargyrite* See page 109.

GRAYISH WHITE **Argentite** See page 67.

GRAYISH WHITE *Jalpaite* See page 67.

GRAYISH WHITE POLYBASITE See page 87.

GALENA WHITE **Stibnite** Sb_2S_3

Microchem. HNO$_3$, Tarnishes dark brown; rubs to gray. HCl, Neg. KCN, Dissolves, bleaches, and roughens surface, but does not tarnish. FeCl$_3$, Neg. HgCl$_2$, Neg. KOH, Tarnishes brown and shows yellow coating; rubs to roughened surface often showing patches of yellow.

Hardness, Low. 2. Slightly brittle. Gray powder when scratched.

Description. C. and Str.—Lead-gray. Crystals are long prisms with striations parallel to elongation. Highly perfect "b" pinacoidal cleavage.

B.B. Yields antimony flame and coat on charcoal (IV, 1, *a*). When pure volatilizes entirely. Fus.—1.

GALENA WHITE KERMESITE Sb_2S_2O

Microchem. HNO$_3$, Tarnishes light brown; rubs pale brown. HCl, Neg. KCN, Slowly tarnishes pale brown; rubs faint. FeCl$_3$, Neg. HgCl$_2$, Neg. (?). KOH, Tarnishes iridescent and is coated with yellow; rubs clean.

Hardness, Low. 1–1.5. Very sectile. Red powder when scratched. Internal reflection as seen by inclined light is red.

Description. C.—Cherry-red. Str.—Brownish-red. Usually in needle-like crystals.

B.B. In the closed tube blackens, fuses, and at first yields a white sublimate of antimony trioxide; after strongly heating yields the usual black or dark red antimony sublimate in closed tube. Otherwise like stibnite. Fus.—1.

GRAYISH WHITE POLYBASITE Ag_9SbS_6

Microchem. HNO$_3$, Some specimens practically negative. Others tarnish very slowly; rubs to faint gray etched surface. HCl, Neg. KCN, Develops scratches and tarnishes dark brown; rubs to gray etched surface. FeCl$_3$, Practically negative, but some specimens tarnish faint brown; wash clean. HgCl$_2$, Quickly tarnishes brown to black; rubs to faint brown. KOH, Neg.

Hardness, Low. 2–3. Sectile, but slightly brittle. Gray powder when scratched. Internal reflection as seen by inclined light is red. (Not always detected.)

Description. C.—Iron-black, in thin splinters cherry-red. Str.—Black.

B.B. On charcoal fuses with spirting to a globule, yielding antimony flame and coat (IV, 1, *a*). Reduced with sodium carbonate yields silver (IV, 18, *a*). Fus.—1.

GALENA WHITE *Livingstonite* See page 111.

WHITE **Chloanthite** $NiAs_2$

Microchem. HNO_3, Slowly tarnishes faint brown; rubs to pale gray. Fumes tarnish. HCl, Neg. KCN, Neg. $FeCl_3$, Slowly tarnishes faint brown, but is often almost negative. $HgCl_2$, Neg. KOH, Neg.
Hardness, High. 5.5–6. Brittle. Gray powder when scratched.
Description. C.—Tin-white. Str.—Gray-black.
B.B. Fuses easily to a globule and yields an arsenic odor and perhaps a white arsenic coat on charcoal. Yields a red precipitate with dimethyl glyoxime (IV, 14, *b*). Gives a cobalt reaction with borax usually due to impurities of that element. Fus.—2.

WHITE *Rammelsbergite* $NiAs_2$ (Very Rare)

Like chloanthite in all properties except crystal system.

WHITE GERSDORFFITE NiAsS

Microchem. HNO_3, Tarnishes brown, then black; rubs to rough gray surface. Fumes tarnish brown and develop structure. HCl, Mineral unaffected, but acid turns bright yellow. KCN, Neg. $FeCl_3$, Slowly tarnishes faint brown. (Sometimes practically negative.) $HgCl_2$, Tarnishes brown; rubs clean. KOH, Neg.
Hardness, 5.5. Brittle.
Description. C.—Silver-white, tarnishing to gray. Str.—Grayish black. Good cubic cleavage.
B.B. Decrepitates in the closed tube and yields a yellowish brown sublimate (IV, 2, *c*). Otherwise like chloanthite.

WHITE *Löllingite* $FeAs_2$ (Very Rare)
Microchem. HNO_3, Slowly tarnished faint brown. Fumes tarnish. HCl, Neg. KCN, Neg. $FeCl_3$, Neg. $HgCl_2$, Neg. KOH, Neg.
Hardness, High. 5–5.5. Brittle.
Description. C.—Silver-white. Str.—Black.
B.B. Fuses to strongly magnetic globule in R.F. on charcoal. Gives arsenic coat and odor; also sublimate and mirror in open and closed tube respectively. Fus.—2.

WHITE *Ulmannite** NiSbS (Very Rare)
Microchem. HNO_3, Tarnishes brown to iridescent; rubs to iridescent gray. HCl, Neg. KCN, Neg. $FeCl_3$, Neg. KOH, Neg.
Hardness, High. 5–5.5. Brittle.
Description. C.—Silver-gray. Str.—Grayish black. Perfect cubic cleavage.
B.B. On charcoal in the R.F. fuses easily to a globule, boils, and yields an antimony coat (IV, 1, *a*). Yields a red precipitate with dimethyl glyoxime (IV, 14, *b*). Fus.—1.5.

CREAMY WHTE *Linnæite* Co_3S_4 (Very Rare)
Microchem. HNO_3, Tarnishes very faint brown. Fumes tarnish brown very slowly. HCl, Neg. KCN, Neg. $FeCl_3$, Neg. $HgCl_2$, Tarnishes iridescent brown. KOH, Neg.
Hardness, High. 5.5. Brittle.
Description. C.—Pale steel-gray, tarnishing copper-red. Str.—Black. Imperfect cubic cleavage.
B.B. On charcoal fuses to a magnetic globule. The roasted mineral yields with the fluxes reactions for nickel, cobalt, and iron usually. Fus.—2.

PINKISH WHITE GLAUCODOT (Co,Fe)AsS (Rare)
Microchem. HNO_3, Slowly tarnishes; rubs to rough gray surface. HCl, Neg. KCN, Neg. $FeCl_3$, Neg. $HgCl_2$, Neg. KOH, Neg.
Hardness, High. 5. Brittle.
Description. C.—Gray. Str.—Black. Usually prismatic crystals or massive.
B.B. Yields arsenic odor and coat on charcoal, etc. (IV, 2). In the R.F. fuses to a feebly magnetic globule, black on the surface, but a light bronze in color when freshly fractured. Fus.—2–3.

CREAM *Hauchecornite* $(Ni,Co)_7(S.Sb.Bi)_8$ (Very Rare)
Microchem. HNO_3, Very slowly tarnishes brown; rubs to very pale brown. Fumes tarnish slightly. HCl, Neg. KCN, Neg. $FeCl_3$, Neg. $HgCl_2$, Slowly tarnishes brown; rubs clean. KOH, Neg.
Hardness, High to medium. 5. Very Brittle. Steel-gray powder when scratched.
Description. C.—Light bronze-yellow. Str.—Black.
B.B. Yields common reactions for constituents indicated in formula. (See Chapter IV.)

WHITE	GERSDORFFITE	See page 89.
WHITE	**Chloanthite**	See page 89.
WHITE	*Willyamite*	See page 57.
CREAMY WHITE	**Pyrite**	See page 57.
CREAM	**Pyrrhotite**	See page 95.

PINK *Breithauptite* NiSb (Very Rare)

Microchem. HNO_3, Blackens; rubs to dark gray. HCl, Neg. KCN, Neg. $FeCl_3$, Tarnishes; rubs to clean-pitted surface. KOH, Neg.

Hardness, Medium to high. 5.5.? Brittle. Reddish-brown powder when scratched.

Description. C.—Light copper-red. Str.—Reddish brown.

B.B. Yields antimony flame and coat on charcoal and with dimethyl glyoxime gives a red precipitate. (IV, 14, *b*). Fus.—1.5-2.

GRAYISH WHITE BOURNONITE $(PbCu_2)3Sb_2S_6$ (Rare)

Microchem. HNO_3, Fumes usually tarnish slowly; rubs clean. HCl, Neg. KCN, Neg. $FeCl_3$, Neg. $HgCl_2$, Tarnishes faintly at edge of drop; rubs to very faint brown. KOH, Neg.

Hardness, Medium to low. 2.5–3. Slightly brittle.

Description. C.—Steel-gray. Str.—Black.

B.B. In the closed tube decrepitates and gives a dark red sublimate. Yields antimony flame and coat on charcoal; with continued heating, yielding a lead coat. Residue reduced with sodium carbonate yields copper buttons. Fus.—1.

GRAYISH WHITE *Stylotypite* $3(Cu_2Ag_2Fe)S.Sb_2S_3$ (Very Rare)

Microchem. HNO_3, Tarnishes brown very slowly; rubs faint or clean. Fumes tarnish. HCl, Neg. KCN, Neg. $FeCl_3$, Neg. $HgCl_2$, Neg. KOH, Neg.

Hardness, Medium. 3. Brittle.

Description. C. and Str.—Iron-black.

B.B. Decrepitates and fuses easily, yielding a steel-gray, magnetic globule. Also tests for constituents in formula. Fus.—1.5.

GRAYISH WHITE *Stannite* $SnCu_2FeS_4$ (Very Rare)

Microchem. HNO_3, Tarnishes brown to iridescent; rubs to very faint brown. HCl, Neg. KCN, Neg. $FeCl_3$, Neg. $HgCl_2$, Neg. KOH, Neg.

Hardness, Medium. 4. Brittle.

Description. C.—Steel-gray. Str.—Black. Massive, granulur, or disseminated.

B.B. In the closed tube decrepitates, giving a faint sublimate only. See Chapter IV for tests for constituents of the mineral.

GRAYISH WHITE *Crookesite** $(CuTlAg)_2Se$ (Very Rare)

Microchem. HNO_3, Slowly tarnishes faint brown; rubs clean. Fumes tarnish. HCl, Neg. KCN, Neg. $FeCl_3$, Neg. KOH, Neg.

Hardness, Medium to low. 2.5–3. Brittle.

Description. C.—Lead-gray. Str.—Black. Massive, compact.

B.B. Fuses easily to greenish black, shining enamel, coloring flame blue. Roasted and reduced with sodium carbonate yields copper and silver. (IV, 6, and 18.) Fus.—1.

GRAYISH WHITE *Hauerite* MnS_2 (Very Rare)

Microchem. HNO_3, Fumes tarnish brown very slowly; washes clean. HCl, Neg. KCN, Neg. $FeCl_3$, Neg. $HgCl_2$, Neg. KOH, Neg.

Hardness, Medium. 4. Brittle. Brick-red powder when scratched.

Description. C.—Brownish black. Str.—Reddish brown.

B.B. Roasted mineral yields green bead with sodium carbonate. Fus.—3.

GRAY **Sphalerite** ZnS

Microchem. HNO_3, Tarnishes faintly brown; rubs clean. HCl, Neg. KCN, Neg. $FeCl_3$, Neg. $HgCl_2$, Neg. KOH, Neg.

Hardness, Medium. 3.5–4. Brittle. Yellow or brown powder when scratched.

Internal reflection as seen by inclined light is yellow or brown.

Description. C.—Shades of yellow, brown to black. Str.—Pale to colorless. Perfect dodecahedral cleavage.

B.B. Nearly infusible. Reduced with soda on charcoal yields white zinc oxide coat, which when moistened with cobalt nitrate sol. and again heated in the O.F. becomes green (IV, 29, *a*).

Wurtzite is like sphalerite physically and microchemically.

YELLOW **Chalcopyrite** See page 117.

CREAMY WHITE *Sulvanite* See page 65.

PINKISH WHITE FAMATINITE See page 83.

GRAY *Voltzite* See page 116.

PALE YELLOW MILLERITE NiS (Rare,
Microchem. HNO_3, Acid is without effect, but fumes tarnish brown; rubs to very faint brown. HCl, Neg. KCN, Neg. $FeCl_3$, Neg. $HgCl_2$, Tarnishes brown; rubs clean. KOH, Neg.
Hardness, Medium. 3–3.5. Brittle. Grayish yellow powder when scratched.
Description. C.—Bronze-yellow. Str.—Greenish black. Commonly occurs as hair-like crystals or botryoidal crusts. At least two perfect cleavages.
B.B. On charcoal in the R.F. the roasted mineral gives a coherent metallic mass, slightly magnetic. As well as yielding reactions for nickel, most varieties also show copper, cobalt, and iron. Fus.—2.

CREAM **Pyrrhotite** FeS(S)x
Microchem. HNO_3, Tarnishes very slowly faint brown; almost negative; rubs clean. Fumes tarnish brown. HCl, Neg. (Fumes sometimes tarnish faintly.) KCN, Neg. $FeCl_3$, Neg. $HgCl_2$, Neg. KOH, Slowly tarnishes brownish or iridescent; rubs to brownish iridescence.
Hardness, Medium. 3.5–4.5. Brittle.
Description. C.—Bronze-yellow. Str.—Grayish-black. Usually massive or disseminated. Naturally magnetic.
B.B. In the R. F. on charcoal blackens and becomes strongly magnetic. Fus.—2.5–3.

CREAM *Chalmersite* $CuFe_2S_3$ (Very Rare)
Microchem. HNO_3, Fumes tarnish slightly; washes clean. HCl, Neg. KCN, Neg. $FeCl_3$, Neg. $HgCl_2$, Neg. KOH, Neg.
Hardness, Medium. 3.5. Slightly brittle.
Description. C.—Pale yellow. Str.—Grayish or greenish black. Strongly, though variably magnetic.
B.B. In the R.F. on charcoal fuses to a strongly magnetic globule. Well roasted and reduced with sodium carbonate on charcoal yields copper (IV, 6, *a*).

CREAMY WHITE PENTLANDITE (Fe,Ni)S (Rare)
Microchem. HNO_3, Slowly tarnishes very faint brown; rubs to very faint brown. HCl, Neg. KCN, Neg. $FeCl_3$, Neg. $HgCl_2$, Neg. KOH, Neg.
Hardness, Medium. 3.5–4. Brittle.
Description. C.—Light bronze-yellow. Str.—Bronze. Non-magnetic. Massive.
B.B. Fuses readily to a magnetic globule on charcoal and yields a red precipitate with dimethyl glyoxime (IV, 14, *b*). Fus.—2.

CREAMY WHITE *Wittichenite* Cu_3BiS_3 (Very Rare)
Microchem. HNO_3, Tarnishes light brown; rubs to gray. Fumes tarnish. HCl, Neg. KCN, Neg. $FeCl_3$, Neg. $HgCl_2$, Neg. KOH, Slowly tarnishes faint brown; rubs clean.
Hardness, 3.5. Slightly brittle.
Description. C.—Steel-gray. Str.—Black.
B.B. On charcoal fuses easily, at first throwing out sparks, and yielding a bismuth oxide coat (IV, 3, *a*). The roasted material, moistened with HCl yields a strong blue copper chloride flame. Fus.—1.

GALENA WHITE *Andorite* $PbAgSb_3S_6$ (Very Rare)
Microchem. HNO_3, Slowly tarnishes brown or iridescent; rubs faint or clean. Fumes tarnish slowly. HCl, Neg. KCN, Neg. $FeCl_3$, Neg. $HgCl_2$, Neg. KOH, Slowly tarnishes to pale brown; rubs to faint brownish iridescence.
Hardness, Medium to low. Brittle.
Description. C.—Steel-gray. Str.—Black.
B.B. Shows lead and antimony as in bournonite, but when cupelled yields silver. Fus.—1.

GRAYISH WHITE **Tennantite** See page 117.

WHITE *Native Antimony* Sb

Microchem. HNO_3, Tarnishes brownish or iridescent; rubs same. HCl, Neg. (Sometimes appears to tarnish very faintly.) KCN, Neg. $FeCl_3$, Tarnishes slowly brown; rubs to pale brown. $HgCl_2$, Slowly tarnishes faintly brown; rubs nearly clean. KOH, Neg.

Hardness, Low. 3.3–5. Very brittle.

Description. C. and Str.—Tin-white. Perfect cleavage.

B.B. On charcoal yields a white coat in both R.F. and O.F.; with intermittent blowing the globule is finally crusted over with prismatic crystals of antimony trioxide.

CREAMY WHITE **Native Silver** See page 43.

GALENA WHITE *Plagionite* See page 43.

GRAYISH WHITE *Aguilarite* See page 76.

WHITE *Berzelianite* Cu_2Se (Very Rare)

Microchem. HNO_3, Tarnishes iridescent; rubs to iridescent greenish gray. HCl, Neg. KCN, Neg. $FeCl_3$, Tarnishes very light brown; rubs clean. $HgCl_2$, Tarnishes very light brown; rubs clean. KOH, Neg.

Hardness, Low. Very sectile.

Description. C.—Silver-white, soon tarnishing. Str.—Shining. Occurs in thin dendritic crusts and disseminated.

B.B. In the open tube gives a red sublimate of selenium, with white crystals of selenium dioxide. On charcoal yields white coat, selenium odor, and blue flame; with soda in R.F. gives metallic copper. Fus.—1.5.

WHITE *Eucairite* $Cu_2Se.Ag_2Se$ (Very Rare)

Microchem. HNO_3, Tarnishes black and shows orange coating; rubs to rough gray. Fumes tarnish slightly. HCl, Neg. KCN, Neg. $FeCl_3$, Practically negative. (Sometimes tarnishes very slowly faint brown.) $HgCl_2$, Neg. KOH, Neg.

Hardness, Low. 2.5. Very sectile.

Description. C.—Silver-gray. Str.—Shining. Massive and granular.

B.B. Like berzelianite except that silver is shown on reduction.

CREAMY WHITE *Kalgoorlite* $HgAu_2Ag_6Te_6$? (Very Rare)

Microchem. HNO_3, Very slowly tarnishes faint brown; rubs almost clean. HCl, Neg. KCN, Neg. $FeCl_3$, Tarnishes iridescent; rubs to rough brownish surface. HgCl, Neg. KOH, Neg.

Hardness, Low. Sectile, but slightly brittle.

Description. C. and Str.—Iron-black. Occurs massive.

B.B. Yields tellurium flame and coat on charcoal (IV, 20, *a*). Roasted and reduced with sodium carbonate yields gold and silver. With soda in a closed tube gives metallic mercury.

GRAYISH WHITE *Petzite* $(Ag, Au)_2Te$ (Very Rare)

Microchem. HNO_3, Tarnishes brown to black; rubs gray. (Some specimens effervesce.) HCl, Neg. KCN, Neg. $FeCl_3$, Tarnishes brown; rubs clean. $HgCl_2$, Tarnishes light brown; rubs faint. KOH, Neg.

Hardness, Low. 2.5–3. Very sectile.

Description. C.—Iron-black. Str.—Gray. Granular to compact.

B.B. Yields a bright green tellurium flame and white coat on charcoal (IV, 20, *a*). Roasted and reduced with soda on charcoal yields gold and silver button. Fus.—1.5.

GRAYISH WHITE *Coloradoite* HgTe (Very Rare)

Microchem. HNO_3, Slowly tarnishes iridescent; rubs clean. HCl, Neg. KCN, Neg. $FeCl_3$, Tarnishes very faint brown; rubs clean. $HgCl_2$, Neg. KOH, Neg.

Hardness, Low. 3. Very sectile.

Description. C.—Iron-black. Str.—Black. Massive; granular.

B.B. Easily volatile on charcoal. In the closed tube slightly decrepitates, fuses and yields metallic mercury, also tellurium oxide in drops, and next to the assay metallic tellurium. Fus.—1.

GRAYISH WHITE **Molybdenite** MoS_2.

Microchem. HNO_3, Slowly very faint tarnish; rubs clean. Fumes tarnish faint. (Reaction is often practically negative.) HCl, Neg. KCN, Neg. $FeCl_3$, Practically negative, but sometimes tarnishes very faint brown; washes clean. $HgCl_2$, Neg. KOH, Neg.

Hardness, Low. 1–1.5. Very sectile.

Description. C.—Lead-gray to bluish black. Str.—Bluish gray. Marks paper. Laminæ are flexible, but inelastic. Perfect basal cleavage.

B.B. Infusible, but yields a green flame, also a white coat in O.F. on charcoal (IV, 13).

GALENA WHITE *Galenobismutite** $PbBi_2S_4$ (Very Rare)

Microchem. HNO_3, Tarnishes brown to black; rubs same. HCl, Neg. KCN, Neg. $FeCl_3$, Neg. KOH, Neg.

Hardness, Low. 3–4.

Description. C.—Lead-gray to tin-white. Str.—Grayish black.

B.B. With KI & S flux shows both lead and bismuth (IV, 3, and 10).

GALENA WHITE *Beegerite** $Pb_6Bi_2S_9$ (Very Rare)

Microchem. HNO_3, Tarnishes; rubs to faint gray. Fumes tarnish. HCl, Neg. KCN, Neg. $FeCl_3$, Neg. KOH, Neg.

Hardness, Low.

Description. C.—Gray. Str.—Grayish black.

B.B. Like galenobismutite.

GALENA WHITE *Freieslebenite* $(Pb,Ag_2)_5Sb_4S_{11}$ (Very Rare)

Microchem. HNO_3, Tarnishes dark; rubs to rough gray. One fairly reliable specimen of this mineral was negative with HNO_3. HCl, Neg. KCN, Neg. $FeCl_3$, Neg. $HgCl_2$, Neg. KOH, Neg.

Hardness, Low. 2–2.5. Brittle.

Description. C.—Steel-gray. Str.—Black. Vertically striated prisms.

B.B. Yields antimony flame and coat on charcoal (IV, 1, *a*). With KI & S flux shows lead (IV, 10, *a*). Reduced with soda and cupelled shows silver. Fus.—1.

GALENA WHITE *Nagyagite* $Au_2Pb_{10}Sb_2Te_6S_{15}$? (Very Rare)

Microchem. HNO_3, Slowly tarnishes iridescent; rubs gray. HCl, Neg. KCN, Neg. $FeCl_3$, Neg. $HgCl_2$, Neg. KOH, Neg.

Hardness, Low. 1–1.5.

Description. C.—Gray. Str.—Black. One perfect cleavage. Thin laminæ are flexible. Tabular crystals; also granular massive.

B.B. Complex, but yields tests for constituents of formula. (See Chap. IV.)

CREAMY WHITE *Sylvanite* $AuAgTe_2$ (Very Rare)

Microchem. HNO_3, Tarnishes brown; rubs to deep brown, sometimes developing cleavage. HCl, Neg. KCN, Neg. $FeCl_3$, Neg. $HgCl_2$, Neg. KOH, Neg.

Hardness, Low. 1.5–2. Slightly brittle.

Description. C.—Yellowish steel-gray. Str.—Gray. Perfect pinacoidal cleavage. Usually crystalline.

B.B. Yields bright green tellurium flame and coat on charcoal (IV, 20, *a*). Roasted and reduced with soda yields both gold and silver. Fus.—1.

CREAMY WHITE *Guanajuatite* Bi_2Se_3 (Very Rare)

Microchem. HNO_3, Quickly tarnishes iridescent; rubs practically clean. Fumes tarnish brown. HCl, Neg. KCN, Neg. $FeCl_3$, Neg. $HgCl_2$, Slowly tarnishes brown; rubs clean. KOH, Neg.

Hardness, Low. 2.5–3.5. Sectile to slightly brittle.

Description. C.—Bluish gray. Str.—Black.

B.B. Yields blue selenium flame, odor, and coat on charcoal (IV, 17, *a*). With KI & S flux yields a brick-red bismuth iodide coat (IV, 3). Fus.—1.5.

PALE BROWN *Sternbergite** $AgFe_2S_3$? (Very Rare)
Microchem. HNO_3, Tarnishes iridescent; rubs gray. HCl, Neg. KCN, Neg. $FeCl_3$, Neg. KOH, Tarnishes brown; rubs clean.
Hardness, Low. 1–1.5.
Description. C.—Bronze. Str.—Black. Perfect basal cleavage. Laminæ are flexible, like tin-foil. Marks paper. Tabular crystals.
B.B. Roasted on charcoal becomes magnetic and the surface of the globule shows separated metallic silver. Fus.—1.5.

GALENA WHITE *Lengenbachite* $Pb_6(AgCu)_2As_4S_{13}$ (Very Rare)
Microchem. HNO_3, Practically negative, but sometimes tarnishes slightly; rubs to faint gray. HCl, Neg. KCN, Neg. $FeCl_3$, Neg. $HgCl_2$, Neg. KOH, Neg.
Hardness, Low. Brittle.
Description. C.—Steel-gray. Str.—Brownish. Marks paper.
B.B. Yields arsenic odor and coat on coal (IV, 2, *a*). See Chapter IV.

GALENA WHITE *Rathite* Pb(As,Sb)S? (Very Rare)
Microchem. HNO_3, Very slowly tarnishes brown; rubs to faint gray. Sometimes practically negative. HCl, Neg. KCN, Neg. $FeCl_3$, Neg. $HgCl_2$, Neg. KOH, Tarnishes brown; rubs to faint gray.
Hardness, Low. Brittle.
Description. C.—Blackish lead-gray. Str.—Reddish brown. Prismatic cryst.
B.B. Yields common tests for constituents in formula. (Chap. IV).

GALENA WHITE *Jordanite** $Pb_4As_2S_7$ (Very Rare)
Microchem. HNO_3, Tarnishes slightly; rubs clean. HCl, Neg. KCN, Neg. $FeCl_3$, Neg. KOH, Neg.
Hardness, Low. 3. Brittle.
Description. C.—Lead-gray. Str.—Black. Six sided crystals.
B.B. Decrepitates strongly yielding arsenic odor and white coat on charcoal; also a yellow lead oxide coat near the assay. Fus.—1.

GALENA WHITE *Guitermanite** $3PbS.As_2S_3$? (Very Rare)
Microchem. HNO_3, Quickly tarnishes iridescent; rubs clean. HCl, Neg. KCN, Neg. $FeCl_3$, Neg. KOH, Neg.
Hardness, Low. 3.
Description. C.—Bluish gray. Str.—Black. Massive, compact.
B.B. Like jordanite.

GALENA WHITE *Epiboulangerite** $Pb_3Sb_2S_8$ (Very Rare)
Microchem. HNO_3, Tarnishes black; rubs to gray. Sometimes shows small white crystals after rubbing. HCl, Neg. (Fumes sometimes tarnish slightly.) KCN, Neg. $FeCl_3$, Neg. KOH, Neg.
Hardness, Low.
Description. C.—Dark bluish gray. Str.—Black. Striated prismatic needles.
B.B. Yields antimony flame and coat on charcoal (IV, 1, *a*); also yellow lead oxide coat near the assay. Fus.—1.

WHITE	*Eucairite*	See page 97.
WHITE	*Berzellianite*	See page 97.
CREAMY WHITE	*Emplectite*	See page 63.
GALENA WHITE	**Stibnite**	See page 87.
GALENA WHITE	*Geocronite*	See page 79.
GRAYISH WHITE	*Crookesite*	See page 94.
GRAYISH WHITE	*Regnolite*	See page 119.
GRAYISH WHITE	JAMESONITE	See page 62.
GALENA WHITE	*Andorite*	See page 95.

GRAYISH WHITE STEPHANITE Ag_5SbS_4

Microchem. HNO_3, Neg. HCl, Fumes slowly tarnish brown to iridescent; rubs clean. KCN, Quickly tarnishes brown; rubs pale brown, developing cracks. $FeCl_3$, Slowly gives a speckled surface; rubs clean. $HgCl_2$, Quickly tarnishes brown; rubs to brown. KOH, Slowly tarnishes deep brown; rubs clean.

Hardness, Low. 2–2.5. Slightly sectile to brittle.

Description. C. and Str.—Iron-black.

B.B. In the closed tube decrepitates and fuses, after long heating giving faint antimony sublimate. On charcoal fuses easily with spirting, yielding a white coat which may become red with oxidized silver. Roasted and reduced with soda yields silver button.

BROWN **Pyrolusite** MnO_2

Microchem. HNO_3, Practically negative. (Faintly tarnishes and dissolves in two or three minutes.) HCl, Tarnishes slowly; rubs pale. KCN, With 20 per cent. solution almost negative, but rapidly tarnishes with very dilute solution and is noticeable usually after washing. $FeCl_3$, Tarnishes dark brown; rubs pale. $HgCl_2$, Neg. KOH, Neg.

Hardness, Low (Variable).

Description. C.—Dark steel-gray to iron-black. Str.—Black. Soils fingers.

B.B. Infusible. Yields manganese reactions with the fluxes. Evolves chlorine when treated with HCl.

GRAYISH WHITE *Jalpaite* See page 67.

BLUISH WHITE **Proustite** See page 109.

GRAYISH WHITE (CREAMY) *Delafossite* CuO, Fe_2O_3? (Very Rare)
Microchem. HNO_3, Neg. HCl, Tarnishes slightly and acid turns yellowish-green; rubs to etched surface. KCN, Neg. $FeCl_3$, Neg. $HgCl_2$, Neg. KOH, Neg.
Hardness, Medium. 2.5. Brittle.
Description. C.—Dark gray. Str.—Blackish gray. Cleavable into laminæ. Occurs in small crystalline plates.
B.B. Fusible with difficulty, coloring the flame green. Easily soluble in HCl.

GRAYISH WHITE POLYBASITE See page 87.

REMOVE
THIS
WRITING

HNO_3-N
HCl-N
KCN
Med.
$FeCl_3$- N

YELLOW	**Chalcopyrite**	See page 117.
BROWNISH WHITE	**Enargite**	See page 83.
GRAYISH WHITE	**PYRARGYRITE**	See page 111.

REMOVE
THIS
WRITING

BLUISH WHITE **Proustite** Ag_3AsS_3

Microchem. HNO_3, Neg. HCl, Neg. Fumes sometimes tarnish faintly; rubs clean. KCN, Slowly tarnishes brown, developing scratches; rubs to faint brown. $FeCl_3$, Slowly tarnishes faintly; rubs clean. $HgCl_2$, Slowly tarnishes brown; rubs to brownish iridescence. KOH, Quickly tarnishes black; rubs to pale brown.

Hardness, Low. 2.5. Slightly brittle. Blood-red powder when scratched.

Internal reflection as seen by inclined light is bright red.

Description.—C. and Str.—Scarlet. Commonly as elongated crystals.

B.B. Yields arsenic odor and white coat on charcoal. Prolonged heating in the O.F. or reduction with soda in the R.F. yields silver. Fus.—1.

BLUISH WHITE *Polyargyrite** $Ag_{24}Sb_2S_{15}$ (Very Rare)

Microchem. HNO_3, Neg. Fumes often tarnish slightly; rubs clean. HCl, Neg. KCN. Quickly tarnishes brown; rubs clean. $FeCl_3$, Tarnishes dark or iridescent; rubs clean. KOH, Neg.

Hardness, Low. 2.5. Very sectile.

Description. C.—Iron-black to grayish black. Str.—Black. Cubic cleavage.

B.B. Fuses easily on charcoal to a black globule, giving antimony flame and coat, and finally a brittle globule of silver. Fus.—1.

GRAYISH WHITE **Cerargyrite** AgCl

Microchem. HNO_3, Neg. HCl, Neg. KCN, Quickly tarnishes to reddish brown; washes to dark solution pit. $FeCl_3$, Quickly tarnishes dark brown; rubs same. $HgCl_2$, Neg. KOH, Slowly tarnishes; rubs pale.

Hardness, Low. 1–1.5. Highly sectile.

Description. C.—Pearl-gray, various. Str.—Waxlike.

B.B. Fuses in closed tube without decomposition. On charcoal gives a globule of metallic silver.

GRAYISH WHITE *Bromyrite* AgBr (Very Rare)

Microchem. HNO_3, Neg. HCl, Neg. KCN, Tarnishes slightly; rubs nearly clean. $FeCl_3$, Tarnishes dark gray to brown; rubs same. $HgCl_2$, Slight tarnish? KOH, Tarnishes rapidly and forms whitish coating.

Hardness, Low. Sectile.

Description. C.—Bright yellow to grass or olive green. Str.—Yellowish green. No cleavage.

B.B. On charcoal emits bromine vapors and yields a globule of metallic silver. In the closed tube reacts like cerargyrite.

Embolite, Ag(Cl, Br), and *iodobromite,* 2AgCl.2AgBr.AgI, are intermediate between cerargyrite, bromyrite and iodyrite in all chemical and physical properties as well as in composition. The differences in color on the polished surface are usually enough to distinguish between any two that may be present together, and the difference in the rates of reaction with any of the active reagents easily brings out contrasts between them.

GRAYISH WHITE *Jalpaite* See page 67.

GRAYISH WHITE *Pearcite* See page 85.

BLUISH WHITE *Miargyrite* $AgSbS_2$ (Very Rare)

Microchem. Like pyrargyrite.

Hardness, Low. 2–2.5. Brittle. Cherry-red powder when scratched.

Internal reflection as seen by inclined light is deep red.

Description. C.—Iron-black to steel-gray. Str.—Cherry-red.

B.B. On charcoal fuses quietly, giving a white antimony coat and a gray bead, which after continued treatment in the O.F. leaves a bright globule of silver. Fus.—1.

GRAYISH WHITE *Argyrodite* Ag_6GeS_6? (Very Rare)

Microchem. HNO_3, Neg. HCl, Neg. KCN, Tarnishes brown, developing scratches; rubs to very faintly etched surface. $FeCl_3$, Neg. $HgCl_2$, Tarnishes brownish iridescent; rubs to iridescence. KOH, Neg.

Hardness, Low. 2.5. Brittle.

Description. C.—Steel-gray on a fresh fracture, with a tinge of reddish violet. Str.—Grayish black.

B.B. In the closed tube yields a brilliant black sublimate; in the open tube fumes of sulphur dioxide. On charcoal fuses to a bead, giving near the assay a faint white sublimate; after long heating, an orange-yellow sublimate and a silver globule. Fus.—1.5.

GRAYISH WHITE ORPIMENT As_2S_3

Microchem. HNO_3, Neg. HCl, Neg. KCN, Tarnishes dark quickly; rubs nearly same. $FeCl_3$, Neg. $HgCl_2$, Yellow coat, rubs clean. KOH, Quickly tarnishes black.

Hardness, Low. 1.5–2. Sectile. Lemon-yellow powder when scratched.

Internal reflection as seen by inclined light is yellow.

Description. C.—Lemon-yellow. Str.—Pale yellow. Usually in foliated or columnar masses. Perfect pinacoidal cleavage.

B.B. On charcoal fuses and volatilizes entirely, yielding arsenic odor. In the closed tube, fuses, volatilizes, and gives a dark yellow sublimate. Fus.—1.

GRAYISH WHITE *Iodyrite* AgI (Very Rare)

Microchem. HNO_3, Neg. HCl, Neg. KCN, Quickly dissolves and tarnishes to dark surface; rubs to gray, etched surface. $FeCl_3$, Neg. $HgCl_2$, Tarnishes brownish to iridescent; rubs same or slightly lighter. KOH, Very slowly yields slight tarnish.

Hardness, Low. Sectile. Internal reflection seen by inclined light is yellowish to brownish.

Description. C.—Yellow to yellowish green, sometimes brownish. Str.—Yellow. Perfect C cleavage.

B.B. In the closed tube fuses and assumes a deep yellow color, but resumes its yellow color on cooling. On charcoal gives fumes of iodine and a globule of metallic silver.

BLUE **Covellite** CuS

Microchem. HNO_3, Neg. HCl, Neg. KCN, Quickly tarnishes purple; rubs to grayish, deeply etched surface. $FeCl_3$, Neg. $HgCl_2$, Neg. KOH, Neg.

Hardness, Low. 1.5–2. Slightly brittle.

Description. C.—Deep indigo-blue. Str.—Gray to black. Perfect basal cleavage; flexible in thin leaves. Common in smaller quantities as a secondary mineral in nearly all copper deposits.

B.B. On charcoal burns with a blue flame, fusing to a globule which reacts like chalcocite (IV, 6). Fus.—2.5.

YELLOW **Native Gold** Au

Microchem. HNO_3, Neg. HCl, Neg. KCN, Slowly blackens; rubs to brownish, rough surface. $FeCl_3$, Neg. (?). $HgCl_2$, Neg. KOH, Neg.

Hardness, Low. 2.5–3. Sectile. Yellow powder when scratched.

Description. C. and Str.—Gold-yellow.

B.B. Fuses easily to yellow button. Not acted on by any of the fluxes.

GALENA WHITE *Livingstonite* $HgSb_4S_7$ (Very Rare)

Microchem. HNO_3, Neg. (?) HCl, Neg. KCN, Slowly tarnishes grayish. $FeCl_3$, Neg. $HgCl_2$, Tarnishes to brownish iridescence; rubs same. KOH, Neg.

Hardness, Low. 2. Red powder when scratched.

Internal reflection as seen by inclined light is deep red.

Description. C.—Lead-gray. Str.—Red. Usually in slender prismatic crystals.

B.B. Roasted on charcoal is completely volatile. Yields antimony flame and coat (IV, 1, *a*). With soda in the closed tube yields mercury (IV, 12, *a*). Fus.—1.

BLUISH WHITE *Onofrite* Hg(SSe) (Very Rare)

Microchem. HNO_3, Neg. HCl, Neg. KCN, Faintly tarnishes brown or bluish, differentially; rubs to faint brown. $FeCl_3$, Neg. $HgCl_2$, Neg. KOH, Neg. Hardness, Low. 2.5. Sectile to slightly brittle. Some flakes show yellow internal reflection as seen by inclined light.

Description. C. and Str.—Blackish gray. Massive; fine granular.

B.B. In the closed tube decrepitates and gives sublimates of sulphur and mercury (IV, 12, *a*). On charcoal gives copious fumes with selenium odor and a sublimate with metallic-like luster which touched by the R.F. disappears, coloring the flame azure-blue.

GRAYISH WHITE PYRARGYRITE Ag_3SbS_3

Microchem. HNO_3, Neg. HCl, Neg. KCN, Tarnishes pale brown slowly; rubs to faint gray surface showing structure. $FeCl_3$, Neg. $HgCl_2$, Slowly tarnishes light brown; rubs clean. KOH, Tarnishes black; rubs to speckled yellowish gray surface.

Hardness, Low. 2.5. Brittle. Blood-red powder when scratched.

Internal reflection as seen by inclined light is brilliant red.

Description. C.—Black to grayish black. Str.—Purplish red.

B.B. On charcoal fuses with spirting to a globule, coats the coal white, and the assay is converted into silver sulphide, which, treated in the O.F., or with soda in the R.F., gives a globule of silver. Fus.—1.

BLUISH WHITE **Proustite** See page 109. (Very Rare)

GRAYISH WHITE POLYBASITE See page 87.

GRAY *Uraninite* Uranate of U,Pb, etc. (Very Rare)

Microchem. HNO_3, Neg. HCl, Neg. KCN, Neg. $FeCl_3$. Slowly tarnishes to still darker brownish gray; rubs same. $HgCl_2$, Neg. KOH, Neg.

Hardness, High. 5.5. Brittle.

Description. C.—Pitch-black to greenish. Str.—Greenish brown to black. Usually massive and botryoidal.

B.B. Infusible. In the R.F. gives a green bead with both borax and salt of phosphorous (IV, 26, *a*). Many impurities are always present and any or all may yield reactions.

Grayish White Wolframite $(FeMn)WO_4$

Microchem. Negative with all reagents.

Hardness, High. 5–5.5. Brittle.

Description. C.—Dark brown to nearly black. Str.—Reddish or brownish to nearly black. Perfect pinacoidal cleavage. Crystals commonly tabular, bladed, columnar, etc.

B.B. Fuses to a magnetic globule with a crystalline surface. With soda yields a green manganate bead (IV, 11, *a*). For tests for tungsten, see IV, 25. Fus.—3.

(Note. Hübnerite, $MnWO_4$, and Ferberite, $FeWO_4$, differ from Wolframite only in color.)

Gray Cassiterite SnO_2

Microchem. Negative with all reagents.

Hardness, High. 6–7.

Description. C.—Brown to black; variable. Str.—Light.

B.B. Infusible. Reduced with soda and borax on charcoal yields metallic tin buttons.

Gray (Variable) **Limonite** $2Fe_2O_3.3H_2O$(?)

Microchem. Negative with all reagents.

Hardness, High. 5–5.5. Brittle. Yellow powder when scratched.

Internal reflection as seen by inclined light is often reddish brown.

Description. C.—Ocher-yellow to brownish black. Str.—Yellowish brown. Never crystallized. Usually botryoidal, concretionary, massive, earthy, etc.

B.B. Infusible. Yields water when heated in the closed tube. Otherwise like hematite.

(Note.—Göthite, $Fe_2O_3.H_2O$, and Turgite, $2Fe_2O_3.H_2O$, behave chemically like Limonite, but Turgite is very much lighter in color.)

Gray Chromite $FeCr_2O_4$

Microchem. Negative with all reagents.

Hardness, High. 5.5.

Description. C.—Iron-black to brownish black; various. Str.—Brown. Commonly massive; fine granular to compact.

B.B. Infusible in O.F., but in R.F. is slightly rounded on edges and becomes magnetic. For tests for chromium see IV, 4.

Grayish White *Rare Earths*

Columbite, Tantalite, Samarskite, etc., are all chemically inert with the reagents used and are indistinguishable on polished sections.

Grayish White Manganese Oxides

Manganite, $Mn_2O_3.H_2O$, Hausmannite, $MnO.Mn_2O_3$, and Braunite, $3Mn_2O_3.MnSiO_3$, are all chemically inert with the reagents used.

WHITE *Sperrylite* $PtAs_2$ (Very Rare)
Microchem. HNO_3, Neg. HCl, Neg. KCN, Neg. $FeCl_3$, Neg. $HgCl_2$, Neg. KOH, Neg.
Hardness, High. 6–7. Brittle.
Description. C.—Tin-white. Str.—Black. In minute isometric crystals.
B.B. Decrepitates slightly. Unchanged in the closed tube. Yields white arsenic trioxide sublimate in the open tube. Dropped on red-hot platinum foil, melts, gives off arsenic trioxide fumes, and deposits porous platinum on the foil. Fus.—2.

PINKISH WHITE **Cobaltite** CoAsS
Microchem. HNO_3, Neg. HCl, Neg. KCN, Neg. (Sometimes slowly faint tarnish rubs clean.) $FeCl_3$, Neg. $HgCl_2$, Neg. KOH, Neg.
Hardness, High. 5.5. Brittle.
Description. C.—Silver-white, inclined to pinkish. Str.—Grayish black. Good cubic cleavage. Often in cubic crystals.
B.B. Unaltered in the closed tube. On charcoal yields an arsenic coat and fuses to a magnetic globule. When well roasted it yields a blue bead with borax. Fus.—2–3.

CREAMY WHITE **Hematite** Fe_2O_3
Microchem. Negative with all reagents.
Hardness, High. 5.5–6.5. Brittle. Brownish red to gray powder when scratched.
Description. C.—Reddish brown, steel-gray, to iron-black. Str.—Cherry-red.
B.B. Infusible. On charcoal in R. F. becomes magnetic.

GRAYISH WHITE **Magnetite** Fe_3O_4
Microchem. Negative with all reagents.
Hardness, High. 5.5–6.5. Brittle.
Description. C.—Iron-black. Str.—Black. Naturally strongly magnetic.
B.B. Nearly infusible. In the O.F. loses its magnetism. With the fluxes reacts like hematite.

GRAYISH WHITE **Ilmenite** $FeTiO_3$
Microchem. Negative with all reagents.
Hardness, High. 5–6. Brittle.
Description. C.—Iron-black. Str.—Black to brownish.
B.B. Infusible in O.F., but slightly rounded on edges in hottest R.F. For wet test for titanium see (IV, 24, *b* and *c*).

GRAYISH WHITE **Rutile** TiO_2
Microchem. Negative with all reagents.
Hardness, High. 6–6.5.
Internal reflection as seen by inclined light sometimes is reddish.
Description. C.—Reddish brown to nearly black. Str.—Pale brown. Occurs usually in prismatic crystals. Knee-like twins.
B.B. Infusible. Yields bead and wet tests for titanium (IV, 24).

GRAYISH WHITE *Franklinite* $(FeZnMn)O.(FeMn)_2O_3$ (Very Rare)
Microchem. Negative with all reagents.
Hardness, High. 5.5–6.5. Brittle.
Description. C.—Iron-black. Str.—Very dark brown.
B.B. Infusible. With soda gives a green manganate bead and on charcoal a coating of zinc oxide (IV, 29).
(Note.—Chalcophanite $(Mn,Zn)O.2MnO_2.2H_2O$, is also grayish white and behaves chemically like Franklinite.)

WHITE *Löllingite* See page 91.

Gray *Voltzite* Zn_5S_4O (Very Rare)

Microchem. Negative with all reagents. (HNO_3, fumes sometimes tarnish very faintly.)

Hardness, Medium. 4–4.5.

Internal reflection as seen by inclined light is reddish yellow or brown.

Description. C.—Dirty rose-red, yellowish, or brownish. Str.—Light. Occurs thin curved, lamellar, etc.

B.B. Nearly infusible. Reduced with soda on charcoal yields white zinc oxide coat, which when moistened with cobalt nitrate solution and again heated in the O.F. becomes green (IV, 29, *a*)

Gray *Erythrozincite** (MnZn)S (Very Rare)

Microchem. Negative with all reagents.

Hardness, Medium to low.

Description. C.—Red. Str.—Pale yellow. Occurs in thin plates.

B.B. Yields green manganate bead with soda. Also usual tests for zinc.

YELLOW **Chalcopyrite** $CuFeS_2$

Microchem. HNO_3, Neg. (Sometimes tarnishes brown; rubs faint. Fumes also tarnish in such cases.) HCl, Neg. KCN, Neg. (Rarely tarnishes iridescent with development of scratches.) $FeCl_3$, Neg. $HgCl_2$, Neg. KOH, Neg.

Hardness, Medium. 3.5–4. Brittle.

Description. C.—Brass-yellow. Str.—Greenish.

B.B. On charcoal fuses to a magnetic globule in R. F. Decrepitates in closed tube and gives sulphur sublimate. Roasted and reduced with soda, yields copper buttons. Fus.—2.

GALENA WHITE *Chalcostibite** $CuSbS_2$ (Very Rare)

Microchem. Negative with all reagents.

Hardness, Medium. 3–4. Brittle.

Description. C.—Blackish gray. Str.—Black. Perfect basal cleavage.

B.B. Yields antimony flame and coat on charcoal. Reduced with soda gives a globule of metallic copper. Fus.—1.5.

GALENA WHITE *Berthierite* $FeSb_2S_4$ (Very Rare)

Microchem. HNO_3, Neg. HCl, Neg. KCN, Neg. $FeCl_3$, Neg. KOH, Slowly tarnishes brownish; rubs faint or clean.

Hardness, Medium to low. 2–3.

Description. C.—Steel-gray. Str.—Black. Elongated prisms or fibrous massive.

B.B. Fuses easily, yielding an antimony flame and coat on charcoal, and a magnetic residue. Fus.—2.

GRAYISH WHITE *Baumhauerite* $Pb_4As_6S_{13}$ (Very Rare)

Microchem. HNO_3, Neg. (Sometimes tarnishes along cracks; rubs clean.) HCl, Neg. KCN, Neg. (Sometimes tarnishes along cracks.) $FeCl_3$, Neg. $HgCl_2$, Slowly tarnishes brown; rubs to pale brown. KOH, Tarnishes iridescent; rubs clean.

Hardness, Medium to low. 3. Brittle.

Internal reflection as seen by inclined light is often red.

Description. C.—Lead-gray. Str.—Brown. One perfect cleavage.

B.B. Yields arsenic odor and coat on charcoal; lead coat near the assay. With KI & S flux gives lemon-yellow lead iodide coat (IV, 10, *a*).

GRAYISH (BROWNISH) WHITE **Tetrahedrite** $Cu_8Sb_2S_7$

Microchem. HNO_3, Neg. (Fumes sometimes tarnish faint brown; rubs clean.) HCl, Neg. KCN, Neg. $FeCl_3$, Neg. $HgCl_2$, Neg. KOH, Neg.

Hardness, Medium. 3–4.5. Very brittle.

Description. C.—Gray to black. Str.—Brown to black.

B.B. Reacts for copper, antimony, arsenic, silver, and often mercury, etc. Almost impossible to distinguish between tetrahedrite and tennantite. Fus.—1.5.

GRAYISH (GREENISH) WHITE **Tennantite** $Cu_8As_2S_7$

Like tetrahedrite into which it grades chemically.

PALE YELLOW	MILLERITE	See page 95.
CREAM	**Pyrrhotite**	See page 95.
CREAM	*Chalmersite*	See page 95.
CREAMY WHITE	PENTLANDITE	See page 95.
GRAYISH WHITE	BOURNONITE	See page 94.

GRAYISH WHITE **Cinnabar** HgS

Microchem. Negative with all reagents used.

Hardness, Low. 2–2.5. Slightly sectile. Carmine-red powder when scratched.

Internal reflection as seen by inclined light is carmine-red.

Description. C.—Cochineal-red, inclining to brownish and lead gray. Perfect prismatic cleavage.

B.B. On charcoal entirely volatile when pure. In closed tube alone gives a black sublimate of mercuric sulphide, but with soda, one of metallic mercury. Fus.—1.5.

GRAYISH WHITE *Metacinnabarite* HgS (Very Rare)

Microchem. Negative with all reagents used. (HNO_3 sometimes tarnishes faintly.)

Hardness, Low. 3. Slightly brittle.

Description. C.—Grayish black. Str.—Black.

B.B. Like cinnabar.

GRAYISH WHITE *Patronite* VS_4(?) (Very Rare)

Microchem. HNO_3, Neg. HCl, Neg. KCN, Neg. $FeCl_3$, Neg. $HgCl_2$, Neg. KOH, Tarnishes iridescent; rubs to paler iridescence.

Hardness. Very low. Sectile.

Description. C.—Bluish black. Str.—Bluish gray. Occurs massive and amorphous.

B.B. Gives test for vanadium with S. Ph. bead. Also wet tests.

GRAY *Lorandite** $TlAsS_2$ (Very Rare)

Microchem. HNO_3, Neg. HCl, Neg. KCN, Neg. $FeCl_3$, Neg. KOH, Quickly develops coating of orange; rubs clean to rough surface.

Hardness, Low. 2–2.5. Dark red powder when scratched.

Internal reflection as seen by inclined light is orange-red.

Description. C.—Carmine-red, often tarnishes gray on the surface. Str.—Cherry-red. Perfect pinacoidal cleavage yielding flexible lamellæ.

B.B. For tests for the rare element thallium see Chapter IV, 21.

GALENA WHITE	*Freieslebenite*	See page 98.
GALENA WHITE	*Berthierite*	See page 117.
GALENA WHITE	*Lengenbachite*	See page 99.
GALENA WHITE	*Rathite*	See page 99.
GRAYISH WHITE	*Baumhauerite*	See page 117.
GRAYISH WHITE	**Molybdenite**	See page 97.
GRAY	*Erythrozincite*	See page 116.

GALENA WHITE *Dufrenoysite* $Pb_2As_2S_5$ (Very Rare)
Microchem. HNO_3, Neg. HCl, Neg. KCN, Neg. $FeCl_3$, Neg. $HgCl_2$, Neg. KOH, Tarnishes iridescent and darkens; rubs to faint gray, etched surface.
Hardness, Low. 3. Brittle.
Description. C.—Blackish gray. Str.—Reddish brown. Perfect basal cleavage.
B.B. On charcoal decrepitates, fuses, and yields a globule of lead which gives trace of silver on cupellation. Also gives arsenic coat on charcoal. Fus.—1.

GALENA WHITE *Matildite** $AgBiS_2$ (Very Rare)
Microchem. Negative with all reagents used.
Hardness, Low.
Description. C.—Gray. Str.—Light gray. Often as slender striated prisms.
B.B. Fuses easily on charcoal, giving a coating of bismuth oxide and on long heating a globule of silver.

GALENA WHITE *Lehrbachite* PbSe + HgSe
Microchem. Negative with all reagents used.
Hardness, Low. Quite brittle.
Description. C.—Lead-gray to iron-black. Occurs massive, granular, etc.
B.B. On charcoal yields to a strong odor of selenium and partly fuses. In the closed tube gives a lustrous metallic gray sublimate of mercury selenide, with sodium carbonate a sublimate consisting of globules of mercury.

BLUISH WHITE *Tiemannite* HgSe (Very Rare)
Microchem. Negative with all reagents used.
Hardness, Low. 2.5. Sectile to slightly brittle.
Description. C.—Blackish gray. Str.—Black.
B.B. Decrepitates in the closed tube and when pure entirely sublimes, giving a black sublimate with the upper edge reddish brown. On charcoal yields selenium odor and blue flame and deposits a white coat with a metallic-like luster.

BLUISH WHITE *Vrbaite** $TlAs_2SbS_5$ (Very Rare)
Microchem. HNO_3, Neg. HCl, Neg. KCN, Neg. $FeCl_3$, Neg. KOH, Tarnishes iridescent; rubs to gray.
Hardness, Low. Internal reflection by inclined light is red. Yellowish red powder when scratched.
Description. C.—Gray-black. Str.—Light yellowish red.

BLUISH WHITE *Stützite** Ag_4Te? (Very Rare)
Microchem. Negative with all reagents used.
Hardness, Low.
Description. C.—Lead-gray with reddish tinge. Str.—Blackish lead-gray.
B.B. Easily fusible to a dark bead from which a silver globule is obtained by reduction with soda. Yields tellurium oxide in the open tube. (IV, 20, *c*).

GRAYISH WHITE *Regnolite* $Cu_7As_2S_{12}$ (Very Rare)
Microchem. Negative with all reagent used. (HNO_3 fumes sometimes tarnish faint brown; rubs clean.)
Hardness, Low.
Resembles tetrahedrite and tennatite very closely.

GRAYISH WHITE *Seligmannite* $CuPbAsS_3$ (Very Rare)
Microchem. HNO_3, Neg. HCl, Neg. KCN, Neg. $FeCl_3$, Neg. $HgCl_2$, Neg. KOH, Slowly tarnishes to iridescence; rubs clean.
Hardness, Low. Sectile to slightly brittle.
Description. C.—Blackish gray. Str.—Brownish black.
B.B. Like bournonite except tests for arsenic instead of antimony.

CHAPTER IV

SUPPLEMENTARY TESTS

TABULATED PROPERTIES OF ORE MINERALS

Some few of the opaque minerals when viewed on a polished surface with vertical illumination show colors other than shades of white or gray. The following table will often be helpful in immediately identifying these minerals.

TABLE 7

Color	Mineral	Formula	Page
Purple	*Umangite*	Cu_3Se_2	77
Purple	*Rickardite*	Cu_4Te_3	35
Blue	**Covellite**	CuS	111
Yellow	**Chalcopyrite**	$CuFeS_2$	117
Yellow	**Gold**	Au	111
Pale yellow	MILLERITE	NiS	95
Cream	(Pyrite)	FeS_2	57
Cream	**(Marcasite)**	FeS_2	57
Cream	**Pyrrhotite**	$FeS(S)_x$	95
Cream	PENTLANDITE	(FeNi)S	95
Cream	*Chalmersite*	$CuFe_2S_3$	95
Cream	*Whitneyite*	Cu_9As	33
Cream	*Maucherite*	Ni_3As_2	55
Cream	*Hauchecornite*	$(NiCo)_7(Bi,S,Sb)_8$	91
Cream	*Sulvanite*	Cu_3VS_4	65
Creamy pink	**Niccolite**	NiAs	55
Coppery pink	**Copper**	Cu	51
Coppery pink	*Breithauptite*	NiSb	93
Brown	**Bornite**	Cu_5FeS_4	53

When the vertical illumination is cut out and the light from some strong source is allowed to strike the polished surface at an inclination, a number of the ore minerals reveal more or less brilliantly colored internal reflections. These colors are quite characteristic of certain minerals and should always be observed in the course of an identification. Table 8 lists those minerals in which such colors are commonly seen.

TABLE 8

Color of internal reflection	Mineral	Formula	Page
Red	*Baumhauerite*	$Pb_4As_6S_{13}$	117
Red	**Cinnabar**	HgS	118
Red	**Cuprite**	Cu_2O	33
Red	*Jalpaite*	$3Ag_2S.Cu_2S$	67
Red	KERMESITE	Sb_2S_2O	87
Red	*Livingstonite*	$HgSb_4S_7$	111
Red	*Miargyrite*	$AgSbS_2$	110
Red	POLYBASITE	Ag_9SbS_6	87
Red	**Proustite**	Ag_3AsS_3	109
Red	PYRARGYRITE	Ag_3SbS_3	111
Red	**(Rutile?)**	TiO_2	115
Red	*Vrbaite*	$TlAs_2SbS_5$	119
Reddish brown	**Limonite**	$2Fe_2O_3.3H_2O$	114
Reddish brown	*Hübnerite*	$MnWO_4$	114
Brown or yellow	(CASSITERITE?)	SnO_2	114
Brown or yellow	**Sphalerite**	ZnS	94
Brown or reddish yellow	*Voltzite*	Zn_5S_4O	116
Brownish or greenish	*Cuprodescloizite*	$(PbZnCu)_4V_2O_9.H_2O$	75
Olive green	ALABANDITE	MnS	41
Orange red	*Lorandite*	$TlAsS_2$	118
Orange	REALGAR	AsS	62
Yellow	(*Onofrite*?)	$Hg(SSe)$	111
Yellow	ORPIMENT	As_2S_3	110

When the polished surface of an ore mineral is deeply scratched or gouged with a sharp needle a powder is usually produced, and although this is gray or black in the majority of minerals, a few yield a powder or scratch with distinct colors as seen under the microscope with vertical illumination. This property may often be used as a valuable aid in mineral identification. The minerals giving powders of characteristic colors in this way are grouped in Table 9 which follows:

TABLE 9

Color of powder	Mineral	Formula	Page
Carmen red	**Cinnabar**	HgS	118
Blood red	**Cuprite**	Cu_2O	33
Blood red	*Jalpaite*	$3Ag_2S.Cu_2S$?	67
Blood red	**Proustite**	Ag_3AsS_3	109
Blood red	PYRARGYRITE	Ag_3SbS_3	111
Red	KERMESITE	Sb_2S_2O	87
Red	*Livingstonite*	$HgSb_4S_7$	111
Red	*Lorandite*	$TlAsS_2$	118
Red	*Miargyrite*	$AgSbS_2$	110
Reddish brown	*Hauerite*	MnS_2	94
Reddish brown	**Hematite**	Fe_2O_3	115
Reddish brown	*Regnolite*	$Cu_7As_2S_{12}$	119
Pink	**Copper**	Cu	51
Golden brown	**Bornite**	Cu_5FeS_4	53
Brown to yellow	**Sphalerite**	ZnS	94
Orange	REALGAR	AsS	62
Reddish yellow	*Cuprodescloizite*	$(PbZnCu)_4V_2O_9.H_2O$	75
Yellow	CHROMITE	$FeCr_2O_4$	114
Yellow	**Gold**	Au	111
Yellow	**Limonite**	$2Fe_2O_3.3H_2O$	114
Yellow	ORPIMENT	As_2S_3	110
Olive-green	ALABANDITE	MnS	41

Table 10.—Classification According to Electrical Conductivity, Measured with Two Dry Cells with Voltmeter, and using Sharp's No. 9 Sewing Needles for Terminals

Electrical conductivity equal to or greater than copper, voltmeter reads 150	Average voltmeter readings	Apparently non-conductors, voltmeter reads 0
Arsenic (native)	147 Onofrite	Aikinite
Algodonite	145 Löllingite	Alabandite
Antimony (native)	142 Petzite	Andorite
Bismuth (native)	142 Bornite	Argyrodite
Calaverite	142 Tiemannite	Baumhauerite
Chalcopyrrhotite	140 Clausthalite	Boulangerite
Chloanthite	137 Arsenopyrite	Bournonite
Coloradorite	137 Eucairite	Brongniardite
Copper (native)	135 Chalcocite	Cassiterite
Covellite	130 Metacinnabarite	Chromite
Dyscrasite	130 Ploydymite	Cinnabar
Gersdorffite	126 Kalgoorlite	Cuprite
Glaucodot	126 Altaite	Cuprodescloisite
Harrisite	124 Chalcopyrite	Cylindrite
Hauchecornite	120 Galena	Delafossite
Hessite	120 Stutzite	Dufrenoysite
Huntilite	120 Luzonite	Emplectite
Kallilite	120 Tennantite	Franklinite
Krennerite	120 Aguilarite	Geocronite
Linnæite	109 Steinmannite	Hauerite
Magnetite	85 Berzellianite	Horsfordite
Maucherite	80 Naumannite	Jalpaite
Melonite	70 Nagyagite	Jordanite
Millerite	57 Polytelite	Kermesite
Mohawkite	55 Stromeyerite	Legenbachite
Niccolite	50 Pearceite	Limonite
Pentlandite	37 Cobaltite	Lorandite
Pyrrhotite	25 Guanajuatite	Matildite
Rammelsbergite	15 Chivatite	Meneghinite
Safflorite	15 Enargite	Miargyrite
Silver (native)	12 Marcasite	Orpiment
Skutterudite	7 Regnolite	Plagionite
Smaltite	7 Lillianite	Polyargyrite
Sternbergite	7 Rezbanyite	Proustite
Sylvanite	7 Pyrite	Pyrargyrite
Tellurium (native)	7 Ilmenite	Rathite
Ullmannite	5 Psilomelane	Realgar
Whitneyite	5 Jamesonite	Rutile
Willyamite	3 Plenargyrite	Seligmannite
	3 Chalmersite	Semseyite
	3 Famatinite	Sphalerite
	3 Cosalite	Stephanite
	2 Bismuthinite	Stibnite
	2 Argentite	Sulvanite
	2 Stannite	Tenorite
	1 Stylotypite	Tetrahedrite
	1 Teallite	Uraninite
	1 Frankeite	Voltzite
	1 Polybasite	Wittichenite
	1 Molybdenite	Wurtzite
	1 Hematite	

TABLE 11.—ELECTRO-POTENTIALS OF ORE MINERALS

Potential Less than 100 per cent. Cu	Between 5 per cent. Au–95 per cent. Cu and 100 per cent. Cu	5 per cent. Au–95 per cent. Cu to 10 per cent. Au–90 per cent. Cu	10 per cent. Au–90 per cent. Cu to 20 per cent. Au–80 per cent Cu	20 per cent. Au–80 per cent Cu to 30 per cent. Au–70 per cent. Cu	30 per cent. Au–70 per cent Cu to 40 per cent Au–60 per cent. Cu	40 per cent. Au–60 per-cent. Cu to 50 per cent Au–50 per cent Cu	50 per cent. Au–50 per cent. Cu to 60 per cent. Au–40 per cent. Cu
Arsenic	Mohawkite	Pyrrhotite	Niccolite	Chloanthite	Linnæite	(Famatinite)	(Pyrite)
	Domeykite	(Domeykite)		Bornite	(Magnetite)	Arsenopyrite	Smaltite
				Chalcocite	Gersdorffite	Millerite	Cobaltite
				Galena	Hessite	Covellite	Pyrolusite
				Tetradymite	Enargite	(Hessite)	
				Dyscrasite	(Pyrite)	Metacinnabarite	
				Magnetite	Famatinite	Coloradoite	
				Marcasite	Löllingite	(Enargite)	
				Nagyagite	Luzonite	Pyrite	
				Native silver	Lehrbachite		
				Stromeyerite	Psilomelane		
				Maucherite			
				Tennantite			
				Tellurium			

TESTS FOR THE ELEMENTS[9]

1. Antimony, Sb

ANTIMONY MINERALS

Minera	Formula	Page
Andorite	$PbAgSb_3S_6$	95
Berthierite	$FeSb_2S_4$	117
Boulangerite	$Pb_3Sb_2S_6$	63
BOURNONITE	$(Cu_2Pb)_3Sb_2S_6$	94
Breithauptite	NiSb	93
Brongniardite	$PbAg_2Sb_2S_5$	69
Chalcostibite	$CuSbS_2$	117
Cylindrite	$Pb_6Sb_2Sn_6S_{21}$	77
Dyscrasite	Ag_6Sb, etc.	67
Epiboulangerite	$Pb_3Sb_2S_8$	99
FAMATINITE	Cu_3SbS_4	83
Franckeite	$Pb_5Sn_2Sb_2S_{12}$	79
Freibergite	$(CuAg)_8Sb_2S_7$	82
Freieslebenite	$(Pb,Ag_2)_5Sb_4S_{11}$	98
Geocronite	$Pb_5Sb_2S_8$	79
Guejarite	$Cu_2Sb_4S_7$	83
Hauchecornite	$(NiCo)_7(BiSbS)_8$	91
Horsfordite	Cu_6Sb	63
JAMESONITE	$Pb_2Sb_2S_5$	62
Kallilite	Ni(SbBi)S	57
KERMESITE	Sb_2S_2O	87
Livingstonite	$HgSb_4S_7$	111
Meneghinite	$Pb_4Sb_2S_7$	79
Miargyrite	$AgSbS_2$	110
Nagyagite	$Au_2Pb_{10}Sb_2Te_6S_{15}$	98
Plagionite	$Pb_5Sb_8S_{17}$	43
Polyargyrite	$Ag_{24}Sb_2S_{15}$	109
POLYBASITE	Ag_9SbS_6	87
PYRARGYRITE	Ag_3SbS_3	111
Rathite	Pb(As,Sb), S	99
Semseyite	$Pb_7Sb_6S_{16}$	45
STEPHANITE	Ag_5SbS_4	101
Stibnite	Sb_2S_3	87
Stylotypite	$3(Cu_2Ag_2Fe)S.Sb_2S_3$	94
Tetrahedrite	$Cu_8Sb_2S_7$	117
Ullmannite	NiSbS	91
Vrbaite	$TlAs_2SbS_5$	119
Willyamite	(CoNi)SbS	57
Zinkenite	$PbSb_2S_4$	62

[9] Reference has been made to standard works on mineralogy and blowpipe analysis by A. H. PHILLIPS, MOSES and PARSONS, BRUSH and PENFIELD, etc.

(*a*) *Coat on Charcoal.*—Antimony minerals, when heated on charcoal in the O.F. yield a white oxide coat which settles fairly near the assay. This coat is volatile and may be completely driven off with either flame. The powdered mineral, when fused with KI & S flux on charcoal, yields a faint yellow coat

(*b*) *Flame Test.*—When the above coat is quickly volatilized in the O.F. the flame is colored a pale yellowish-green. (Arsenic coat colors the flame faint violet.)

(*c*) *Closed Tube Reaction.*—Most minerals containing both antimony and sulphur, when heated in the closed tube, yield a black sublimate when hot, which becomes reddish-brown upon cooling.

(*d*) *Open Tube Reaction.*—Minerals containing antimony and sulphur, when heated in the open tube, yield dense white fumes which settle along the lower half of the tube. This sublimate is volatile and can be entirely driven off.

2. Arsenic, As

ARSENIC MINERALS

Mineral	Formula	Page
Arsenic	As	59
Arsenopyrite	$FeAsS$	57
Baumhauerite	$Pb_4As_6S_{13}$	117
Chloanthite	$NiAs_2$	89
Cobaltite	$CoAsS$	115
Domeykite	Cu_3As	35
Dufrenoysite	$Pb_2As_2S_5$	119
Enargite	Cu_3AsS_4	83
GERSDORFFITE	$NiAsS$	89
GLAUCODOT	$(CoFe)AsS$	91
Guitermanite	$Pb_3As_2S_6$	99
Jordanite	$Pb_4As_2S_7$	99
Lengenbachite	$Pb_6(AgCu)_2As_4S_{13}$	99
Löllingite	$FeAs_2$	91
Lorandite	$TlAsS_2$	118
Luzonite	Cu_3AsS_4	83
Maucherite	Ni_3As_2	55
Niccolite	$NiAs$	55
ORPIMENT	As_2S_3	110
Pearcite	Ag_9AsS_6	85
Proustite	Ag_3AsS_3	109
Rammelsbergite	$NiAs_2$	89
Rathite	$Pb(AsSb)S$	99
REALGAR	AsS	62

Mineral	Formula	Page
Regnolite	$Cu_7As_2S_{12}$	119
Safflorite	$CoAs_2$	55
Seligmannite	$CuPbAsS_3$	119
Smaltite	$CoAs_2$	55
Sperrylite	$PtAs_2$	115
Tennantite	$Cu_8As_2S_7$	117
Vrbaite	$TlAs_2SbS_5$	119
Whitneyite	Cu_9As	33

(*a*) *Coat on Charcoal.*—Arsenic minerals, when heated on charcoal in the O.F. yield a white oxide coat which settles at a distance from the assay. (Beyond the antimony coat.) This coat is extremely volatile in either flame and, with the R.F. especially, strong characteristic garlic-like odors are given off.

(*b*) *Flame Test.*—When arsenic minerals are heated on charcoal in the R.F. the flame is colored a faint violet. (Antimony yields a pale, yellowish green flame.)

(*c*) *Closed Tube Reaction.*—Fusible arsenic minerals mixed with coal dust are placed in the bottom of a narrow closed tube and covered with a small splinter of charcoal. First heat the coal to glowing and then the mineral and coal dust mixture. Any arsenic present, reduced in passing over the hot carbon, deposits as a mirror on the cold walls of the tube. The mirror may be tested by breaking off the bottom of the tube and heating in the Bunsen flame. The characteristic garlic-like odor of arsenic can now be detected as the fumes escape from the mouth of the tube. These fumes also impart a faint violet tinge to the flame if allowed to escape in it. Most arsenic minerals when heated in the closed tube without a flux yield a sublimate which is dark red when hot and reddish-yellow when cold.

(*d*) *Open Tube Reactions.*—Arsenic minerals, when *very slowly* heated in the open tube, yield a ring of white crystals which are very volatile and may be completely driven off from the tube upon heating. No arsenic odor is usually noticed in this test.

3. Bismuth, Bi

BISMUTH MINERALS

Mineral	Formula	Page
Aikinite	$PbCuBiS_3$	63
Beegerite	$Pb_6Bi_2S_9$	98
Bismuth	Bi	43
Bismuthinite	Bi_2S_3	62
Chilenite	Ag_6Bi	35
Chiviatite	$Pb_2Bi_2S_{11}$	63
Cosalite	$Pb_2Bi_2S_5$	59
Dognacskaite	Cu, Bi, and S	63
Emplectite	$CuBiS_2$	63
Guanajuatite	Bi_2Se_3	98
Hauchecornite	$(NiCo)_7(BiSSb)_8$	91
Kallilite	$Ni(SbBi)S$	57
Lillianite	$Pb_3Bi_2S_6$	77
Matildite	$AgBiS_2$	119
Rezbanyite	$Pb_4Bi_{10}S_{19}$	61
Tapalpaite	$3Ag_2(STe).Bi_2(STe)_3$	43
Tetradymite	$Bi_2(Te,S)_3$	61
Wittichenite	Cu_3BiS_3	93

(*a*) *Coat on Charcoal.*—Bismuth minerals, finely powdered and mixed with 2 or 3 parts of sodium carbonate, when heated in the R.F., yield metallic globules which are brittle instead of malleable (as in the case of lead). If the heating is continued, oxide forms and volatilizes, depositing a yellow coat close to the assay. This coat is very similar to that of lead and is best distinguished from it by means of the iodide reaction. The above coat moistened with hydriodic acid and heated gently in the O.F., or the powdered mineral mixed with 2 or 3 parts of KI & S flux and similarly heated, yields a brick-red bismuth iodide coat at some distance from the assay, inside of which there is often a yellow oxide coat.

4. Chromium, Cr

CHROMITE	$FeCr_2O_4$	Page 114

(*a*) *Bead Tests.*—The bead tests for chromium, as for most other elements are only satisfactory when other substances do not interfere. The colors of the borax and S.Ph. beads are given in Table 17, page 149.

(*b*) *Wet Test.*—Fuse the finely ground chromium mineral mixed with 4 parts sodium carbonate and 2 parts of potassium nitrate on a platinum wire. The fusion is dissolved in 2 or 3 c.c. of water, acidified with acetic acid, and filtered if the solution is not clear. If 2 or 3 drops of lead acetate are now added, any chromium present will be precipitated as yellow lead chromate. This may be filtered off and tested as in (*a*).

5. Cobalt, Co

COBALT MINERALS

Minera	Formula	Page
Cobaltite	$CoAsS$	115
GLAUCODOT	$(CoFe)AsS$	91
Hauchecornite	$(NiCo)_7(BiSSb)_8$	91
Linnæite	Co_3S_4	91
Safflorite	$CoAs_2$	55
Smaltite	$CoAs_2$	55
Willyamite	$(CoNi)SbS$	57

(*a*) *Bead Tests.*—Cobalt minerals, when dissolved in S.Ph. or borax, after roasting on charcoal, yield a deep blue bead in all flames. This test is quite sensitive, but if very large amounts of copper or nickel are present the borax bead should be taken from the wire and reduced beside tin on charcoal. The copper and nickel are absorbed by the tin and the bead will remain blue.

6. Copper, Cu

COPPER MINERALS

Mineral	Formula	Page
Aikinite	$PbCuBiS_3$	63
Berzelianite	Cu_2Se	97
Bornite	Cu_5FeS_4	53
BOURNONITE	$(PbCu_2)_3Sb_2S_6$	94
Chalcocite	Cu_2S	51
Chalcopyrite	$CuFeS_2$	117
Chalcostibite	$CuSbS_2$	117
Chalmersite	$CuFe_2S_3$	95
Copper	Cu	51
Covellite	CuS	111
Crookesite	$(CuTlAg)_2Se$	94
Cuprite	Cu_2O	33
Cuprodescloizite	$(PbZnCu)_4V_2O_9.H_2O$	75
Delafossite	CuO,Fe_2O_3	103
Dognacskaite	Cu, Bi, and S	63
Domeykite	Cu_3As	35
Emplectite	$CuBiS_2$	63
Enargite	Cu_3AsS_4	83

Mineral	Formula	Page
Eucairite	$Cu_2Se.Ag_2Se$	97
FAMATINITE	Cu_3SbS_4	83
Freibergite	$(CuAg)_3Sb_2S_7$	82
Guejarite	$Cu_2Sb_4S_7$	83
Horsfordite	Cu_6Sb	63
Jalpaite	$Cu_2S.3Ag_2S$	67
Lengenbachite	$Pb_6(AgCu)_2As_4S_{13}$	99
Luzonite	Cu_3AsS_4	83
Regnolite	$Cu_7As_2S_{12}$	119
Rickardite	Cu_4Te_3	35
Seligmannite	$CuPbAsS_3$	119
Stannite	$SnCu_2FeS_4$	94
Stromeyerite	$(Ag,Cu)_2S$	67
Stylotypite	$3(Cu_2Ag_2Fe)S.Sb_2S_3$	94
Sulvanite	Cu_3VS_4	65
Tennantite	$Cu_8As_2S_7$	117
TENORITE	CuO	73
Tetrahedrite	$Cu_8Sb_2S_7$	117
Umangite	Cu_3Se_2	77
Whitneyite	Cu_9As	33
Wittichenite	Cu_3BiS_3	95

(*a*) *Reduction on Charcoal.*—Copper minerals, roasted in O.F. on charcoal, then finely ground and mixed with two or three parts of sodium carbonate and borax, when treated in a strong R.F. on charcoal, yield malleable copper buttons. It may sometimes be necessary to crush and wash the charge in the mortar to obtain the buttons if small.

(*b*) *Flame Tests.*—Copper minerals heated in O.F. on charcoal; moistened with HCl and then reheated in the R.F., yield a brilliant azure-blue copper chloride flame, which may be tinged with the green flame of copper oxide. This azure-blue flame is very sensitive and will detect a fraction of 1 per cent. of copper. Copper oxide and a few copper minerals when powdered and heated directly in the O.F. yield an emerald-green flame.

(*c*) *Bead Tests.*—Copper minerals, when roasted in O.F. on charcoal, and dissolved in S.Ph. or borax, yield a green bead while hot, which becomes blue when cold. In the R.F. if much oxide is present, the cold bead in reflected light is an opaque red due to cuprous oxide, Cu_2O.

(*d*) *Wet Test.*—Copper salts color all acid solutions blue or green. With an excess of ammonium hydroxide the solution turns deep blue. Iron, if present, may be held in solution by adding tartaric acid before introducing the ammonia.

7. Germanium, Ge

Argyrodite	Ag_6GeS_5	Page 110

(*a*) *Coat on Charcoal.*—Argyrodite, when heated in the R.F. and O.F. on charcoal, yields a glazy white coat near the assay, which assumes a yellowish color at a distance from it.

8. Gold, Au

GOLD MINERALS

Mineral	Formula	Page
CALAVERITE	$AgAuTe_2$	61
Gold	Au	111
Kalgoorlite	$HgAu_2Ag_6Te_6$	97
KRENNERITE	$AgAuTe_2$	63
Nagyagite	$Au_2Pb_{10}Sb_2Te_6S_{15}$	98
Petzite	$(AgAu)_2Te$	97
Sylvanite	$AgAuTe_2$	98

Gold is usually present in such small amounts that ordinary blowpipe methods are not of much value in detecting it. It is present as a major constituent only in the tellurides, and reducing these minerals with sodium carbonate on charcoal in the R.F. yields a malleable button containing gold and silver. After dissolving with nitric acid the black residue is ignited and turns gold yellow.

9. Iron, Fe

IRON MINERALS

Mineral	Formula	Page
Arsenopyrite	FeAsS	57
Berthierite	$FeSb_2S_4$	117
Bornite	Cu_5FeS_4	53
Chalcopyrite	$CuFeS_2$	117
Chalmersite	$CuFe_2S_3$	95
CHROMITE	$FeCr_2O_4$	114
Delafossite	CuO,Fe_2O_3	103
Ferberite	$FeWO_4$	114
GLAUCODOT	(CoFe)AsS	91
Göthite	$Fe_2O_3.H_2O$	114
Franklinite	$(FeZnMn)O.(FeMn)_2O_3$	115
Hematite	Fe_2O_3	115
Ilmenite	$FeTiO_3$	115
Limonite	$2Fe_2O_3.3H_2O$	114

Mineral	Formula	Page
Lollingite	$FeAs_2$	91
Magnetite	Fe_3O_4	115
Marcasite	FeS_2	57
PENTLANDITE	(FeNi)S	95
Pyrite	FeS_2	57
Pyrrhotite	$FeS(S)_x$	95
Stannite	$SnCu_2FeS_4$	94
Sternbergite	$AgFe_2S_3$	99
Stylotypite	$3(Cu_2Ag_2Fe)S.Sb_2S_3$	94
Turgite	$2Fe_2O_3.H_2O$	114
WOLFRAMITE	$(FeMn)WO_4$	114

(*a*) *Magnetic Test.*—Iron minerals when treated on charcoal in the R.F. become magnetic after cooling. Cobalt and Nickel also become magnetic when so treated and, when their presence is suspected, may be tested for as on pages 129 and 135.

(*b*) *Bead Tests.*—The colors obtained upon dissolving iron minerals after roasting in S.Ph. or borax are not very satisfactory and should not be depended upon alone for identifications. For these tests see Table 17, page 149.

(*c*) *Wet Test.*—When ammonium hydroxide is added in excess to a solution obtained by boiling any iron mineral in nitric acid, a heavy precipitate of reddish-brown ferric hydroxide is thrown down. In this way iron may be quantitatively separated from the solution. The precipitate filtered off may be tested in the S.Ph. or borax bead or it may be ignited on charcoal and tested for magnetism.

10. Lead, Pb

LEAD MINERALS

Mineral	Formula	Page
Aikinite	$PbCuBiS_3$	63
Altaite	PbTe	43
Andorite	$PbAgSb_3S_6$	95
Baumhauerite	$Pb_4As_6S_{13}$	117
Beegerite	$Pb_6Bi_2S_9$	98
Boulangerite	$Pb_3Sb_2S_6$	63
BOURNONITE	$(PbCu_2)_3Sb_2S_6$	94
Brongniardite	$PbAg_2Sb_2S_5$	69
Chiviatite	$Pb_2Bi_2S_{11}$	63
Clausthalite	PbSe	77
Cosalite	$Pb_2Bi_2S_5$	59
Cuprodescloizite	$(PbZnCu)_4V_2O_9.H_2O$	75

Mineral	Formula	Page
Cylindrite	$Pb_6Sb_2Sn_6S_{21}$	77
Dufrenoysite	$Pb_2As_2S_5$	119
Epiboulangerite	$Pb_3Sb_2S_8$	99
Franckeite	$Pb_5Sn_2Sb_2S_{12}$	79
Freieslebenite	$(PbAg_2)_5Sb_4S_{11}$	98
Galena	PbS	77
Geocronite	$Pb_5Sb_2S_8$	79
Guitermanite	$Pb_3As_2S_6$	99
JAMESONITE	$Pb_2Sb_2S_5$	62
Jordanite	$Pb_4As_2S_7$	99
Lehrbachite	PbSe + HgSe	119
Lengenbachite	$Pb_6(AgCu)_2As_4S_{13}$	99
Lillianite	$Pb_3Bi_2S_6$	77
Meneghinite	$Pb_4Sb_2S_7$	79
Nagyagite	$Au_2Pb_{10}Sb_2Te_6S_{15}$	98
Naumannite	$(Ag_2Pb)Se$	43
Plagionite	$Pb_5Sb_8S_{17}$	43
Rathite	Pb(As,Sb),S	99
Rezbanyite	$Pb_4Bi_{10}S_{19}$	61
Seligmannite	$CuPbAsS_3$	119
Semseyite	$Pb_7Sb_6S_{16}$	45
Teallite	$PbSnS_2$	77
Uraninite	Uranate of U, Pb, Th, etc.	113
Zinkenite	$PbSb_2S_4$	62

(*a*) *Coat on Charcoal.*—Lead minerals when slowly heated on charcoal yield a volatile yellow oxide coat which deposits close to the assay. (If much sulphur is present and the heating is rapid, a white coat similar to the antimony coat is formed—this does not occur when the heating is done slowly.) This yellow coat, moistened with hydriodic acid and heated gently in the O.F., or the powdered mineral mixed with 2 or 3 parts of KI & S flux and similarly heated, yields a lemon-yellow lead iodide coat. (Similarly treated, Bismuth yields a brick-red coat.)

(*b*) *Reduction on Charcoal.*—Lead minerals powdered and roasted in a very small O.F., then mixed with 4 parts sodium carbonate, 1 part borax, and 1 part charcoal dust, and treated with the R.F. on charcoal, yield soft, highly malleable metallic lead buttons. (Bismuth buttons are brittle.)

(*c*) *Wet Test.*—Sulphuric acid added to a nitric acid solution of lead minerals, brings down a white powdery precipitate of lead sulphate. This, when filtered off, may be tested on charcoal as in (*a*) and (*b*).

11. Manganese, Mn

MANGANESE MINERALS

Mineral	Formula	Page
ALABANDITE	MnS	41
BRAUNITE	$3Mn_2O_3.MnSiO_3$	114
Erythrozincite	$(MnZn)S$	116
Franklinite	$(FeZnMn)O.(FeMn)_2O_3$	115
Hauerite	MnS_2	94
HAUSMANNITE	$MnO.Mn_2O_3$	114
Hübnerite	$MnWO_4$	114
MANGANITE	$Mn_2O_3.H_2O$	114
PSILOMELANE	H_4MnO_5	71
Pyrolusite	MnO_2	101
WOLFRAMITE	$(FeMn)WO_4$	114

(*a*) *Sodium Carbonate and Niter Bead.*—Any manganese mineral and any mineral containing as much as 0.1 per cent. of manganese will give a decisive test with this method, which, moreover, is not interfered with by the presence of any other substances. The finely powdered mineral, after roasting on charcoal, is fused with sodium carbonate on a platinum wire in the O.F. and again fused with a small grain of potassium nitrate in the O.F. The resulting bead will be green to dark blue, depending upon the amount of manganese present.

(*b*) *Common Bead Tests.*—Manganese imparts characteristic colors to the borax bead, and to a lesser extent, to the S.Ph. bead. See Table 17, page 149, for these colors.

12. Mercury, Hg

MERCURY MINERALS

Mineral	Formula	Page
Cinnabar	HgS	118
Coloradoite	$HgTe$	97
Kalgoorlite	$HgAu_2Ag_6Te_6$	97
Lehrbachite	$PbSe + HgSe$	119
Livingstonite	$HgSb_4S_7$	111
Metacinnabarite	HgS	118
Onofrite	$Hg(SSe)$	111
Tiemannite	$HgSe$	119

(*a*) *Closed Tube Reaction.*—Mercury minerals, powdered and mixed with 3 parts sodium carbonate, placed in a closed tube with a little sodium carbonate on top of the charge, and heated, yield a gray sublimate of metallic mercury globules on the walls of the tube. The common mercury mineral, cinnabar (and metacinnabarite) when heated alone in the closed tube, yields a black sublimate on the cold walls of the tube.

(*b*) *Open Tube Reaction.*—Mercury minerals, when heated very slowly in the open tube, yield a gray sublimate of metallic mercury.

13. Molybdenum, Mo

Molybdenite MoS_2

(*a*) *Coat on Charcoal.*—Molybdenite, strongly heated in the O.F. on charcoal, yields a volatile oxide coat which is yellowish while hot and white when cold. If this coat is touched instantaneously with the R.F. it turns deep blue. Close about the assay a non-volatile film of copper-red MoO_2 may sometimes be seen.

(*b*) *Bead Tests.*—After roasting on charcoal, molybdenite imparts characteristic colors to the S.Ph. and borax beads. See Table 17, on page 149.

14. Nickel, Ni

NICKEL MINERALS

Mineral	Formula	Page
Briethauptite	$NiSb$	93
Chloanthite	$NiAs_2$	89
GERSDORFFITE	$NiAsS$	89
Hauchecornite	$(NiCo)_7(BiSbS)_8$	91
Kallilite	$Ni(SbBi)S$	57
Maucherite	Ni_3As_2	55
Melonite	Ni_2Te_3	61
MILLERITE	NiS	95
Niccolite	$NiAs$	55
PENTLANDITE	$(FeNi)S$	95
Polydymite	Ni_4S_5	57
Rammelsbergite	$NiAs_2$	89
Ullmannite	$NiSbS$	91
Willyamite	$(CoNi)SbS$	57

(*a*) *Bead Tests.*—For the colors imparted by nickel to the borax and S.Ph. beads see Table 17, page 149. Iron, cobalt, and copper, etc. will often entirely obscure the nickel color. It may often be detected in the presence of these elements by the following procedure: The mineral is fused to a globule and most of the arsenic, antimony, or sulphur roasted off. A piece of borax twice the size of the globule is placed beside it on the charcoal and heated in the O.F. for a short time, and the color of the bead observed. Iron, cobalt, nickel, and copper oxidize in the order named, and if successive charges of borax are each treated briefly in the O.F. beside the globule, each element will in turn impart its color to the bead. The amount of each present may be roughly estimated from the number of fresh charges of borax required to absorb it. For example, if it takes two or three beads to remove the cobalt it is present only as a minor constituent and after the cobalt blue has disappeared, the borax will be colored reddish brown by any nickel present. Even with this procedure the results are not always satisfactory, in which case the following wet reaction must be resorted to.

(*b*) *Wet Test.*—The powdered mineral is dissolved in 4 or 5 c.c. of nitric acid, an equal amount of tartaric acid is added, and ammonium hydroxide to excess. The tartaric acid holds the iron in solution. The liquid may be filtered if not clear and then made neutral or very slightly acid with HCl. If 3 or 4 drops of a 1 per cent. solution of dimethyl glyoxime in alcohol are now added, any nickel present will be thrown down as a brilliant red precipitate. This precipitate, when filtered off, may be tested for nickel in the beads as in (*a*). This test is very sensitive and is not interfered with by other elements provided the iron is held in solution by sufficient tartaric acid.

15. Oxygen, O

Oxygen Minerals

Mineral	Formula	Page
Cassiterite	SnO_2	114
Chromite	$FeCr_2O_4$	114
Cuprite	Cu_2O	33
Cuprodescloizite	$(PbZnCu)_4V_2O_9.H_2O$	75
Delafossite	$CuO.Fe_2O_3$	103
Franklinite	$(FeZnMn)O.(FeMn)_2O_3$	115
Ferberite	$FeWO_4$	114
Hematite	Fe_2O_3	115
Hübnerite	$MnWO_4$	114
Ilmenite	$FeTiO_3$	115
Kermesite	Sb_2S_2O	87
Limonite	$2Fe_2O_3.3H_2O$	114
Magnetite	Fe_3O_4	115
Psilomelane	H_4MnO_5	71
Pyrolusite	MnO_2	101
Rutile	TiO_2	115
Tenorite	CuO	73
Uranninite	Uranate of U, Pb, Th, etc.	113
Voltzite	Zn_5S_4O	116
Wolframite	$(FeMn)WO_4$	114

Oxygen usually cannot be directly tested for in the minerals, but its presence is inferred from a knowledge of the character and behavior before the blowpipe of the other constituents.

16. Platinum, Pt

NATIVE PLATINUM Pt *Sperrylite* $PtAs_2$

Platinum is usually recognized from its physical properties and insolubility in acids. If present it is collected in the reduction and cupellation for silver and gold. (See page 139.) If the button so obtained, or the residue after parting with nitric acid, is dissolved in aqua regia, evaporated nearly to dryness, a little hydrochloric acid added, and again almost evaporated, diluted with a little water, and concentrated ammonium chloride added, a precipitate of yellow ammonium platinic chloride is thrown down. Platinum tests are usually delicate operations and the standard references on qualitative analysis should be consulted.

17. Selenium, Se

SELENIUM MINERALS

Mineral	Formula	Page
Aguilarite	$Ag_2S.Ag_2Se$	76
Berzelianite	Cu_2Se	97
Clausthalite	PbSe	77
Crookesite	$(CuTlAg)_2Se$	94
Eucairite	$Cu_2Se.Ag_2Se$	97
Guanajuatite	Bi_2Se_3	98
Lehrbachite	PbSe + HgSe	119
Naumannite	$(Ag_2Pb)Se$	43
Onofrite	Hg(SSe)	111
Tiemannite	HgSe	119
Umangite	Cu_3Se_2	77

(*a*) *Coat on Charcoal.*—Selenium minerals, when heated on charcoal, yield a white oxide coat with a metallic-like luster, sometimes reddish at the edges.

(*b*) *Flame Test.*—When the above coat is heated in the R.F. the flame is colored an intense azure-blue and a disagreeable characteristic odor is given off. This odor has been likened to that of horse radish, but really must be experienced in order to be recognized. It is noticed when even very small amounts of selenium are present.

(c) *Closed Tube Reaction.*—Some selenium minerals, when heated in the closed tube, yield fused brownish red globules of selenium on the walls of the tube.

(d) *Open Tube Reaction.*—Selenium minerals, when heated in the open tube, yield colorless globules of selenous oxide, which crystallize and whiten when cold. The characteristic azure-blue color will be obtained if the volatile oxide fumes are allowed to escape into the Bunsen flame.

18. Silver, Ag

SILVER MINERALS

Mineral	Formula	Page
Aguilarite	$Ag_2S.Ag_2Se$	76
Andorite	$PbAgSb_3S_6$	95
Argentite	Ag_2S	67
Argyrodite	Ag_6GeS_5	110
Brongniardite	$PbAg_2Sb_2S_5$	69
CALAVERITE	$AgAuTe_2$	61
Cerargyrite	$AgCl$	109
Chilenite	Ag_6Bi	35
Crookesite	$(CuTlAg)_2Se$	94
Dyscrasite	Ag_6Sb, etc.	67
Eucairite	$Cu_2Se.Ag_2Se$	97
Freieslebenite	$(PbAg_2)_5Sb_4S_{11}$	98
Freibergite	$(CuAg)_8Sb_2S_7$	82
HESSITE	Ag_2Te	66
Jalpaite	$3Ag_2S.Cu_2S$	67
Kalgoorlite	$HgAu_2Ag_6Te_6$	97
Lengenbachite	$Pb_6(AgCu)_2As_4S_{13}$	99
Matildite	$AgBiS_2$	119
Miargyrite	$AgSbS_2$	110
Naumannite	$(Ag_2Pb)Se$	43
Pearcite	Ag_9AsS_6	85
Petzite	$(AgAu)_2Te$	97
Polyargyrite	$Ag_{24}Sb_2S_{15}$	109
POLYBASITE	Ag_9SbS_6	86
Proustite	Ag_3AsS_3	109
PYRARGYRITE	Ag_3SbS_3	111
Silver	Ag	43
STEPHANITE	Ag_5SbS_4	101
Sternbergite	$AgFe_2S_3$	99
Stromeyerite	$(AgCu)_2S$	67
Stützite	Ag_4Te	119
Stylotypite	$3(Cu_2Ag_2Fe)S.Sb_2S_3$	94
Sylvanite	$AuAgTe_2$	98

(*a*) *Reduction on Charcoal.*—Silver minerals, when powdered and well roasted in O.F., then mixed with 3 or 4 parts of sodium carbonate and fused on charcoal in the R.F., yield metallic silver, which should be collected as much as possible into one button. If the button is not bright it should be heated beside borax on charcoal in the O.F. The base metals are oxidized first and dissolve in the borax, leaving a bright, malleable silver button, which may be further tested by dissolving in nitric acid and precipitating the silver as white silver chloride by adding hydrochloric acid. This silver chloride precipitate, when filtered off is insoluble in hot water, while a similar lead precipitate is dissolved by boiling water. Mercurous mercury also yields a white chloride precipitate which is blackened by ammonium hydroxide, but mercury should not be found in a button treated as in the above case.

When the suspected silver mineral occurs in very small amount and cannot be separated, conclusive tests can be applied providing the microscope reveals no other mineral likely to contain silver. The powdered sample is mixed with an equal volume each of pure test lead and borax glass, and fused and reduced by the R.F. in a deep cavity on charcoal. Manipulate the assay to collect the lead in a single button. The O.F. is now applied to oxidize impurities of arsenic, antimony, etc., and continued until the globule boils freely. It is now allowed to cool and is freed of slag by hammering into a small cube. A cupel is now prepared by compacting dry bone ash in a cavity in charcoal about a centimeter in diameter and one-half centimeter deep. The agate pestle is conveniently used in this operation and by finishing up with a twirling motion the cupel will have a hard, smooth, concave surface. Any loose bone ash is blown off and the cupel is strongly ignited in the O.F. to drive off all moisture, The button is now placed on the cupel and fused in the R.F. for a few moments; then the O.F. is applied and continued without interruption until all the lead is oxidized and absorbed by the bone ash, the end point being marked by a distinct change in color and brightening of the button. If very little silver is present the bead will appear to disappear entirely, but the spot should always be carefully examined with the lens. The residual button will contain any silver, gold, or metals of the platinum group present in the sample. If the bead is treated with a few drops of nitric acid and hydrochloric acid added to the liquid, any silver present will be thrown down as insoluble white silver chloride.

19. Sulphur, S

Sulphur is present in most metallic ores and tests for the element seldom aid determinations in mineragraphic work.

(*a*) *Roasting on Charcoal.*—Sulphur in sulphides, heated on charcoal in the O.F., yields sulphur dioxide fumes which are recognized by their odor.

(*b*) *Sodium Carbonate Test.*—Powdered sulphide, fused with about 3 parts of sodium carbonate, when placed on a bright silver coin and moistened with a little water, stain the coin brown or black. (Since very faint reactions are sometimes due to the sulphur in the gas used, the fusion may be made in a closed tube in cases of doubt.)

20. Tellurium, Te

TELLURIUM MINERALS

Mineral	Formula	Page
Altaite	$PbTe$	43
CALAVERITE	$AgAuTe_2$	61
Coloradoite	$HgTe$	97
HESSITE	Ag_2Te	66
Kalgoorlite	$HgAu_2Ag_6Te_6$	97
KRENNERITE	$AgAuTe_2$	63
Melonite	Ni_2Te_3	61
Nagyagite	$Au_2Pb_{10}Sb_2Te_6S_{15}$	98
Petzite	$(AgAu)_2Te$	97
Rickardite	Cu_4Te_3	35
Stützite	Ag_4Te	119
Sylvanite	$AuAgTe_2$	98
Tellurium	Te	61
Tetradymite	$Bi_2(Te,S)_3$	61

(*a*) *Coat on Charcoal.*—Tellurium minerals, when heated on charcoal in the R.F., yield a white oxide coat resembling the antimony coat, which also colors the flame pale greenish.

(*b*) *Closed Tube Reaction.*—Tellurium minerals, fused in the closed tube with 3 parts of sodium carbonate and some charcoal dust, when dissolved in water after cooling, yield a reddish violet colored solution which gives a gray tellurium precipitate if poured out and exposed to the air.

(*c*) *Open Tube Reaction.*—Tellurium minerals, when heated in the open tube, yield heavy white oxide fumes which condense close to the heated portion of the tube. This sublimate is volatilized with difficulty and fuses to yellow globules which become colorless on cooling.

(*d*) *Sulphuric Acid Test.*—Tellurium minerals, powdered and heated in a test tube with 2 or 3 c.c. of concentrated sulphuric acid, yield a reddish violet solution which gives a precipitate as in (*b*) when cooled and diluted with a little water.

21. Thallium, Tl

Thallium Minerals

Mineral	Formula	Page
Crookesite	$(CuTlAg)_2Se$	94
Lorandite	$TlAsS_2$	118
Vrbaite	$TlAs_2SbS_5$	119

(*a*) *Coat on Charcoal and Flame.*—Thallium minerals, when heated on charcoal in the R.F., yield a white oxide coat and color the flame bright green. This coat moistened with hydriodic acid and heated gently in the O.F., or the powdered mineral mixed with 2 or 3 parts of KI & S flux and similarly heated, yields a lemon-yellow coat very similar to that of lead. The bright green flame, however, will serve to distinguish it from lead.

22. Thorium and the Rare Earths

These rare elements which occur in uraninite, columbite, tantalite, samarskite, etc., are detected by somewhat complicated wet methods. For their application reference should be had to complete works on analytical chemistry.

23. Tin, Sn

Tin Minerals

Mineral	Formula	Page
Cassiterite	SnO_2	114
Cylindrite	$Pb_6Sb_2Sn_6S_{21}$	77
Franckeite	$Pb_5Sb_2Sn_2S_{12}$	79
Stannite	$SnCu_2FeS_4$	94
Teallite	$PbSnS_2$	77

(*a*) *Reduction Test and Coat on Charcoal.*—Tin minerals, powdered and roasted, then mixed with 4 parts sodium carbonate, 1 part borax, and 1 part charcoal dust, and treated with the R.F.

on charcoal, yield white malleable metallic tin globules which give a white insoluble precipitate on heating in a test tube with nitric acid. (The charge should not be heated long after reduction as metallic tin is easily volatile.) An oxide coat is usually formed very near the assay. This coat, which is yellow while hot and white when cold, is very like the zinc coat, and like it, is volatilized with difficulty. If the tin coat is moistened with cobalt nitrate solution and heated in the O.F. it turns blue or bluish green when cold. (A zinc coat similarly treated is grass green.)

24. Titanium, Ti

TITANIUM MINERALS

Mineral	Formula	Page
Ilmenite	$FeTiO_3$	115
Rutile	TiO_2	115

(*a*) *Bead Tests.*—The bead reactions for titanium are not very decisive and are interfered with by other elements. They will be found in Table 17, page 149.

(*b*) *Wet Test with Metallic Tin.*—A charge containing not less than 3 per cent. titanium may be treated as follows: The finely powdered mineral mixed with 6 parts of sodium carbonate and a little borax is strongly fused on charcoal. The fusion is dissolved in 2 or 3 c.c. of concentrated hydrochloric acid, granulated tin added and the solution heated. The liquid takes on a violet color especially upon standing a few minutes.

(*c*) *Hydrogen Peroxide Test.*—For charges containing only a very small amount of titanium the fusion is carried out as directed in the foregoing paragraph, but is boiled in a test tube with 2 or 3 c.c. of dilute sulphuric acid. When dissolved, dilute with about 10 c.c. water and add about 2 c.c. hydrogen peroxide, which turns the solution yellow to orange depending on the amount of titanium present.

25. Tungsten, W

TUNGSTEN MINERALS

Mineral	Formula	Page
Ferberite	$FeWO_4$	114
Hübnerite	$MnWO_4$	114
Tungstenite	WS_2	62
WOLFRAMITE	$(FeMn)WO_4$	114

(*a*) *Bead Tests.*—The bead colors for tungsten will be found in Table 17, page 149.

(*b*) *Wet Test.*—Tungstates, dissolved in the S.Ph. bead, then reduced beside tin on charcoal, powdered, and boiled with 1 or 2 c.c. of dilute hydrochloric acid and a little granulated tin, yield a characteristic blue solution.

26. Uranium, U

Uraninite	Uranate of U, Pb, Th, etc.	Page 113

(*a*) *Bead Tests.*—The colors which uranium gives when dissolved in the S.Ph. bead usually serve to identify it. See Table 17, page 149.

(*b*) *Wet Test.*—The powdered mineral is fused with sodium carbonate and dissolved in hydrochloric acid, nearly neutralized with ammonium hydroxide, a solution of sodium carbonate added until precipitation is complete, then half as much more. Let stand and filter. Acidify the filtrate with hydrochloric acid, boil to expel CO_2, and add an excess of ammonium hydroxide. The precipitate of yellow ammonium urinate may be filtered off and tested in the S.Ph. bead as in (*a*).

27. Vanadium, V

Vanadium Minerals

Mineral	Formula	Page
Cuprodescloizite	$(PbZnCu)_4V_2O_9.H_2O$	75
Patronite	VS_4	118
Sulvanite	Cu_3VS_4	65

(*a*) *Bead Tests.*—Vanadium minerals yield characteristic colors with the beads, especially S.Ph. See Table 17, page 149.

(*b*) *Wet Test.*—The well roasted vanadium mineral is fused with 4 parts of sodium carbonate and 2 parts of potassium nitrate, the fusion is powdered and dissolved in boiling water and the insoluble residue filtered off. The alkali vanadate formed will be found in the filtrate, which is acidified with acetic acid and lead acetate added. A light yellow precipitate of lead vanadate is formed and turns white on standing. (Lead chromate so formed is of a brighter yellow color.) The precipitate may be filtered off and tested in the S.Ph. bead as in (*a*).

28. Water, H_2O

MINERALS YIELDING WATER

Mineral	Formula	Page
Cuprodescloizite	$(PbZnCu)_4V_2O_9.H_2O$	75
Gothite	$Fe_2O_3.H_2O$	114
Limonite	$2Fe_2O_3.3H_2O$	114
PSILOMELANE	H_4MnO_5	71
Turgite	$2Fe_2O_3.H_2O$	114

When a powdered hydrated mineral is heated in the closed tube, water of crystallization is expelled and condenses upon the cold walls of the tube.

29. Zinc, Zn

ZINC MINERALS

Mineral	Formula	Page
Cuprodescloizite	$(PbZnCu)_4V_2O_9.H_2O$	75
Erythrozincite	$(ZnMn)S$	116
Franklinite	$(FeZnMn)O.(FeM)_2O_3$	115
Sphalerite	ZnS	94
Voltzite	Zn_5S_4O	116
Wurtzite	ZnS	94

(*a*) *Coat on Charcoal.*—Zinc minerals, when intensely heated in the R.F. on charcoal, yield a zinc oxide coat which deposits close to the assay. This coat is pale yellow while hot and white when cold. It is non-volatile in the O.F. and difficultly volatile in the R.F. When the powdered zinc mineral is fused with an equal part of sodium carbonate and a little charcoal dust the coat is usually heavier and more satisfactory. This coat, if moistened with cobalt nitrate solution and heated in the O.F. turns grass-green, at least in spots. (The tin oxide coat similarly treated is blue or bluish green.)

TABLES OF IMPORTANT CHEMICAL AND BLOWPIPE REACTIONS

Table 12.—Microchemical Reactions[10]

[10] From W. L. Whitehead's condensation of the complete tables of microchemical tests on the opaque minerals found in the second volume of "Behrens-Kley Mikrochemische Analyse," Kley, P. D. C., 1915, pp. 109–130

Metal	Reagent	Precipitate	Remarks
Arsenic	Silver nitrate	Lemon yellow	Solution in HCl. Arsenic must be as arsenite
Antimony.	Cesium chloride	Hexagonal plates	
Bismuth	Cesium chloride	Rhombic plates	
Cobalt	Ammonium mercuric sulphocyanate	Dark blue prisms	Neutral or slightly acid (acetic) solution
Copper	Potassium ferrocyanide	Amorphous red brown	Acetic acid solution
Gold	Thallium nitrate	Citron yellow needles	Solution must be strong and neutral
Iron	Potassium ferrocyanide	Dark blue sol in alk	Much Cu interferes
Lead	Potassium iodide	Bright yellow hex plates	Slightly acid (nitric) solution
Manganese	Potassium chromate	Yellow brown crystals	Solution neutral
Mercury	Potassium iodide	Mercuric-vermilion Mercurous-amorphous bright yellow	Reagent added solid
Nickel	Dimethyl glyoxime	Acicular magenta colored	In ammoniacal solution
Selenium	Potassium iodide	Powdery brown red	HCl solution of selenates
Silver	Hydrochloric acid	Amorphous white soluble in	Nitric acid solution
Sulphur	Calcium acetate	White needles of $CaSO_4$	Nitric acid solutions of sulphides give sulphates
Tellurium	Cesium chloride	Citron yellow octahedra	HCl solution of tellurium dioxide
Tin	Cesium chloride	Colorless octahedra and cubes	Interfering elements numerous
Titanium	Rubidium chloride	Hex and octagonal plates	Reagent added to solution of sodium flurotitanate
Uranium.	Sodium acetate	Tetrahedrons light yellow	Weak acetic acid solution
Vanadium	Silver nitrate	Pointed yellow grains	Acid solution pyrovanadate
Zinc	Ammonium mercuric sulphocyanate	White feathery crystals in crosses and aggregates	

Table 13.—Standard Fusibilities

Fusibility	Standard	Character of fusion
1	Stibnite	Large fragments fuse in the yellow flame
2	Chalcopyrite	Small fragments fuse in the yellow flame
3	Pyrite	Coarse fragments globular in the O.F.
4	(Actinolite)	Coarse edges rounded in the O.F. No opaque representatives of this class
5	(Orthoclase)	Needle-like fragments become globular in the O.F. A few nearly infusible opaque minerals belong here

TABLE 14.—HEATING ON CHARCOAL
(With or without fluxes)

Sublimate, reduction, etc	Element	Remarks
White sublimate at a considerable distance from the assay. Very volatile	*Arsenic*	Often yields strong garlic odor
White sublimate nearer the assay than the above. Volatile	*Antimony*	
White sublimate with metallic-like luster. Very volatile. Sometimes reddish at the edges	*Selenium*	The sublimate touched with the R.F. imparts an azure-blue color to the flame
White sublimate like antimony. Volatile	*Tellurium*	The sublimate touched with the R.F. volatilizes and colors the flame green
Sublimate near the assay which is pale yellow when hot, white when cold. Not volatile in the O.F.	*Zinc*	This sublimate moistened with cobalt nitrate solution and ignited becomes green
Sublimate which is pale yellow when hot, white when cold. Copper-red close to the assay	*Molybdenum*	The sublimate if touched instantaneously with the R.F. becomes azure-blue
Sublimate which is faint yellow to white when hot, white when cold. Not volatile in the O.F.	*Tin*	The sublimate moistened with cobalt nitrate solution and ignited becomes bluish-green
Sublimate which is yellow when hot, pale yellow when cold. Volatile in the O.F. and R.F.	(*Sulphur* and *Lead*)	This sublimate resembles the antimony coat and forms when galena or other lead sulphides are heated very quickly and intensely
Sublimate close to assay which is deep yellow when hot, pale yellow when cold. Volatile in O.F. and R.F.	*Lead*	This sublimate moistened with HI and ignited forms lemon-yellow lead iodide coat
Sublimate which is deep yellow when hot, pale yellow when cold. Volatile in the O.F. and R.F.	*Bismuth*	This sublimate moistened with HI and ignited forms a brick-red bismuth iodide coat
Reduction with soda yields brittle gray button as well as coat	*Antimony*	

TABLE 14.—*Continued*

Sublimate, reduction, etc.	Element	Remarks
Reduction with soda yields gray infusible particles as well as coat	*Molybdenum*	
Reduction with soda yields malleable white button as well as coat.	*Tin*	
Reduction with soda yields gray malleable button as well as coat	*Lead*	
Reduction with soda yields reddish white brittle button as well as coat	*Bismuth*	
Reduction with soda yields malleable buttons, but no coat	*Copper*, *Silver*, or *Gold*	
Reduction with soda yields gray magnetic particles, but no coat	*Iron*, *Cobalt*, or *Nickel*	

TABLE 15.—SUBLIMATES IN THE CLOSED TUBE

Nature of sublimate	Element	Remarks
Mirror-like; collects in globules	*Mercury*	Mercury minerals fused with sodium carbonate
Black; turns red when rubbed	*Mercury*	Cinnabar alone in closed tube
Mirror-like; does not collect in globules	*Arsenic* and *Tellurium*	Arsenic minerals fused with charcoal dust
Deep red when hot; reddish yellow when cold	*Arsenic*	The sulpharsenides heated alone in closed tube
Black when hot; reddish brown when cold	*Antimony*	Some sulphantimonides alone in closed tube
Red when hot; yellow, cold. White if in small amount	*Sulphur*	Many sulphides
Red to black; becomes red when rubbed.	*Selenium*	Selenides

TABLE 16.—REACTIONS IN THE OPEN TUBE

Nature of sublimate, etc.	Element	Remarks
Dense fumes, depositing white powder mostly on under side of the tube. Pale yellow while hot, white when cold. Non-volatile and infusible	*Antimony*	All sulphantimonides
White, volatile, and crystalline sublimate	*Arsenic*	Sulpharsenides, etc.
White, non-volatile sublimate, fusible to colorless or pale yellow globules	*Tellurium*	
White, non-volatile sublimate, fusible to yellow globules; white when cold	*Lead*	Sulphides containing lead
White, non-volatile sublimate which is infusible	*Bismuth*	Sulphides containing bismuth
White, volatile, and crystalline sublimate. Often shows some red at distance from the assay	*Selenium*	
Crystalline sublimate near the assay; yellow while hot, white when cold	*Molybdenum*	
Volatile metallic mirror of gray globules	*Mercury*	Globules will unite when rubbed

TABLE 17.—BEAD REACTIONS WITH BORAX AND SALT OF PHOSPHORUS

Borax		Salt of phosphorus		Element
O F	R.F.	O F	R F	
Hot, yellow to red; cold, yellow to colorless	Green	Hot, yellow to red; cold, yellow to colorless	Dirty brownish or greenish	Iron
Hot, yellow; cold, colorless	Brown	Yellowish green to colorless	Fine green	Molybdenum
Hot, pale yellow; cold, colorless	Yellow to brown	Yellow to colorless	Hot, dirty blue; cold, fine blue	Tungsten
Hot, pale yellow; cold, colorless	Grayish to brown	Pale yellow to colorless	Hot, yellow; cold, violet	Titanium
Hot, yellow; cold, pale yellow	Colorless	Hot, yellow; cold, colorless	Colorless	Cerium
Hot, yellow, cold, yellowish green	Fine green	Fine green	Fine green	Chromium
Hot, yellow; cold, yellow to colorless	Fine green	Yellow	Fine green	Vanadium
Hot, orange, cold, yellow	Green	Greenish yellow	Fine green	Uranium
Hot, green; cold, blue	Colorless to opaque red	Green to blue	Colorless to opaque red	Copper
Blue	Blue	Blue	Blue	Cobalt
Hot, violet; cold, reddish brown	Opaque gray	Reddish yellow	Reddish yellow	Nickel
Hot, violet, cold, reddish violet	Colorless	Violet	Colorless	Manganese

INDEX

Printed in Dunstable, United Kingdom

91226245R00094